Mechanics
IUTAM
USNC/TAM

Carl T. Herakovich

Mechanics IUTAM USNC/TAM

A History of People, Events, and Communities

 Springer

Carl T. Herakovich
University of Virginia
Charlottesville, VA, USA

ISBN 978-3-319-81252-6   ISBN 978-3-319-32312-1   (eBook)
DOI 10.1007/978-3-319-32312-1

Printed on acid-free paper

This Springer imprint is published by Springer Nature
The registered company is Springer International Publishing AG Switzerland

In memory:
Jim Simmonds
(1935–2015)
Professor of Applied Mathematics
and Mechanics
at Thomas Jefferson's University.
In the true Jefferson mold, Jim exemplified
leadership, scholarship, and citizenship.
He was a dear friend and colleague.

Photo Credits

The great majority of photos presented in the book were available in the public domain. In many cases, permission was given by the individual to use their photograph. A few of the photos required citation.

- Bernoulli—painting by J. R. Huber
- Cauchy—Lithography after painting by Jean Roller
- D'Alembert—1763 portrait by Emanuel Handmann
- Den Hartog—MIT
- Drucker—Courtesy of the University of Illinois Archives, RS 39/2/20
- Ehrenfest—portrait—Dibner Library of the History of Science and Technology—Smithsonian
- Einstein—photo by Ferdinand Schmutzer
- Euler—1763 portrait by Emanuel Handmann
- Hetenyi—Society for Experimental Mechanics
- Hooke—found by Lisa Jardine, Royal Society
- Kirchhoff—U.S. Library of Congress
- Koiter—1993 photo by Monique Simmonds
- Lagrange—engraving by Robert Hart
- Laplace—portrait by Jean-Baptiste Paulin Guérin
- Lord Rayleigh—Royal Society of London
- Millikan—Getty images
- Muskhelishvili—Oil by G. Totibadze
- Navier—bust at École Nationale des Ponts et Chaussees
- Newton—1689 Portrait by Godfrey Kneller
- Poisson—lithograph by François Delpech
- Prager—by William C. Klenk, Brown University Portrait Collection
- Prandtl—Portrait by unknown author, DLR Archive, Gottingen
- Reynolds—1904 painting by John Collier
- Southwell—portrait by Henry Lamb, Imperial College

- St. Venant—National Library of France
- Timoshenko—Timoshenko Collection, Stanford Library, Stanford, California
- von Kármán—The Wind and Beyond: Theodore von Kármán, Little, Brown and Company
- von Mises—Konrad Jacobs photo, Creative Commons, Share Alike 2.0 Germany

Preface

This book reviews the history of the people, events, and communities that led to the development and growth of *Mechanics*, the International Union of Theoretical and Applied Mechanics (*IUTAM*), and the United States National Committee on Theoretical and Applied Mechanics (*USNC/TAM*). Initially, I planned to write a book that was a factual history of the USNC/TAM. However, it quickly became apparent that the people involved in the development of the science of mechanics, from ancient time to the present, are vital to any such history. Also, understanding the development of IUTAM is critical to understanding the establishment of the USNC/TAM.

My experience includes 20 years on the USNC/TAM, including 12 as secretary (2000–2012) and 18 years as a member of the IUTAM General Assembly (1998–2016). The years as secretary of the USNC/TAM made me realize that the early history of the committee (prior to 1982) was not well documented. Philip G. Hodge, Jr., who was secretary for the 18 years (1982–2000) preceding my term as secretary did an outstanding job of documenting the information available to him. He also established procedures and policies for maintaining historical records of the committee. Unfortunately, other than what Hodge maintained, it does not appear that a detailed record of the committee activities has survived the test of time. Detailed minutes of meetings are available for the period since the beginning of Hodge's term in 1982, but little prior to that time. I have expanded on the history before 1982 using newly released data from IUTAM.

The book provides a history of the people who have made significant contributions to the development of the science of mechanics, those who developed IUTAM, and those who led the establishment of the USNC/TAM. It serves as a permanent repository giving details of events leading up to the formation of IUTAM (1922–1924), the establishment of the USNC/TAM (1946–1948), and the leaders and activities of the USNC/TAM during the following 65 years. Special attention is given to the people: who were they, where they came from, and what they contributed.

It became clear to me while researching the book that the people who played important roles in mechanics were, and are, both outstanding scientists/engineers and interesting personalities. At times, it was difficult not to delve in the details of their lives. Some of this will be evident in the following pages. I have been fortunate to personally know 29 of the 32 members of the USNC/TAM who served as chair of the committee or a leadership position in IUTAM.

One of the issues I faced when writing was the question: to what extent should I include references to the many outstanding scholarly works of the individuals discussed. It was clear that some limits had to be established. I decided to include references to only those works that I considered to be classic or more pertinent to this history. Books and other contributions with a substantial amount of mechanics history are provided as a separate listing. These works provide many detailed listings of additional references.

The information provided in this history is gleaned from the books with history, web sites, available minutes of USNC/TAM meetings, personal experiences, and the individuals themselves if they were living at the time of writing about them. Unfortunately, several have died since the book was started. Numerous people have assisted me in this endeavor and I have provided a separate listing of them. All living persons reviewed a draft write-up about them and approved the final version, including permission to use their photo.

This book is not a history of *all* developments in mechanics nor *all* contributors to the field during the period covered. In particular, numerous individuals who have made very significant contributions to mechanics, but were not involved directly with the leadership of the USNC/TAM or IUTAM, are not discussed in this book.

I want to give special thanks to Professor Mike Hyer at Virginia Tech who reviewed the entire manuscript and gave me many helpful suggestions.

Most importantly I would to thank my wife Marlene for more than fifty-five years of love and support.

Charlottesville, VA   Carl T. Herakovich
December 2015

"On the Shoulders of Giants"

Sir Isaac Newton, 1676

Acknowledgements

Writing a history book that encompasses a large number of people and events over many centuries requires the assistance of many people and organizations, as well as the review of many books, articles, and other writings. Listed below are individuals and organizations who either provided direct assistance or served as resource material.

- American Society of Mechanical Engineers, New York, NY
- Anthony D. Rosato, New Jersey Institute of Technology, Newark, NJ
- Caltech—oral histories library
- Dan Frederick, Virginia Tech, Blacksburg, Virginia
- Dave J. Soukup, ASME, New York, NY
- Encyclopedia Britannica
- Giovanni Battimelli—University of Rome, Rome, Italy
- Janice F. Goldblum, National Academy of Sciences Archives, Washington, DC
- MacTutor History of Mathematics archive, University of St. Andrews, Scotland
- Mike Hyer, Virginia Tech, Blacksburg, Virginia
- Monique and Jim Simmonds, University of Virginia, Charlottesville, Virginia
- Norm Abramson, Southwest Research Institute, San Antonio, Texas
- Paola Agnola, CISM, Udine, Italy
- Physics of Fluids, College Park, Maryland
- Sarah Lester, Timoshenko Collection, Stanford Library, Stanford, CA
- Susan Tolbert, Smithsonian, Washington, DC
- The Royal Society, London
- U.S. National Academies, Washington, DC
- U.S. National Aeronautics and Space Administration, Washington, DC

Abbreviations

USNC/TAM Societies

AAM	American Academy of Mechanics
AIAA	American Institute of Aeronautics and Astronautics
AIChE	American Institute for Chemical Engineering
AMS	American Mathematical Society
APS	American Physical Society
ASA	Acoustical Society of America
ASCE	American Society of Civil Engineers
ASME	American Society of Mechanical Engineers
ASTM	American Society for Testing and Materials
SEM	Society for Experimental Mechanics
SES	Society of Engineering Science, Inc.
SIAM	Society for Industrial and Applied Mathematics
SNAME	Society for Naval Architects and Marine Engineers
SOR	Society of Rheology
USACM	United States Association for Computational Mechanics

Academies, Committees, and Organizations

AFOSR	Air Force Office of Scientific Research
ACS	American Chemical Society
Caltech	California Institute of Technology
GALCIT	Guggenheim Aeronautical Laboratory California Institute of Technology
IACM	International Association for Computational Mechanics
ICC	International Congress Committee
IUTAM	International Union of Theoretical and Applied Mechanics

JPL	Jet Propulsion Laboratory
JSCES	Japan Society for Computational Engineering and Science
JSME	Japan Society of Mechanical Engineers
MAL	Member-at-Large
NACA	National Advisory Committee for Aeronautics
NAE	National Academy of Engineering
NAS	National Academy of Sciences
NASA	National Aeronautics and Space Administration
NBS	National Bureau of Standards
NRC	National Research Council
NSF	National Science Foundation
ONR	Office of Naval Research
SCORDIM	SubCommittee On Research Directions In Mechanics
USNC/TAM	U.S. National Committee on Theoretical and Applied Mechanics

Contents

Chapter 1
Mechanics: An Engineering Science

Aristotle often is credited with the first written use of the word *mechanics* in his fourth century BC treatise *Mechanics (Greek Mechanae)*, also translated as *Mechanical Problems*. The ideas presented in this manuscript are Aristotelian in content; however, there is considerable dispute as to whom actually wrote the treatise (Winter 2007; Dugas 1988).

Present-day use of the term *mechanics* by the public often refers to a variety of *"how things work"*. Examples include the lexicon of sports terminology, *mechanics* of the golf swing, *mechanics* of the throwing* motion, *mechanics* of tackling, as well as artisans as *mechanics*, e.g., *auto mechanic*. Your author has spent an untold number of hours over more than 25 years studying the *"mechanics of football officiating"*. The number of examples of this usage of *mechanics* is limitless.

While these applications of the term *mechanics* are valid in that they describe, *"how things work"*, they are not the focus of this book. As used in this book, *mechanics* refers to *why things work the way they do* as well as *how things work*. Further, *theoretical mechanics* refers to mathematical formulations that describe, or model, the phenomena; *experimental mechanics* refers to the test methods used to verify that the theoretical formulations are in agreement with observations in the laboratory or real-world. *Applied mechanics* refers to the application of fundamental mechanics principles for the solution of real-world problems. The term *theoretical and applied mechanics* covers all aspects of the mathematical formulation, experimental verification, and application for understanding physical phenomena. There is virtually no limit to the physical phenomena under considered.

As described by Beer and Johnston (1962) in their text book, "Mechanics may be defined as that science which describes and predicts the conditions of rest or motion of bodies under the action of forces. It is divided into three parts: mechanics of rigid bodies, mechanics of deformable bodies, and mechanics of fluids." Note that *fluids* include both gases and liquids. It is also important to recognize that Beer and Johnston define mechanics as a *science*.

© Springer International Publishing Switzerland 2016
C.T. Herakovich, *Mechanics IUTAM USNC/TAM*,
DOI 10.1007/978-3-319-32312-1_1

Mechanics is a science that combines mathematics, physics, and engineering to provide mathematical statements that describe the behavior of materials and structures. Early mechanicians typically were mathematicians or physicists who applied their discipline to provide rational explanations for observed physical phenomena, i.e., to explain the *mechanics of how things work*. The science of mechanics has grown exponentially over the past century. Mechanics now serves as the fundamental bases for understanding and predicting physical phenomena ranging from body functions at the nanoscale to the vehicles and travel paths required for space exploration.

Another term that frequently appears in mechanics is *rational mechanics*. Truesdell (1977) in his book *A First Course in Rational Continuum Mechanics* (2nd edition), defines *Rational Mechanics* as: "*the part of mathematics that provides and develops logical models for the enforced changes of place and shape we see everyday things suffer*". He states further "The things mechanics represents by mathematical constructs include animals and plants, mountains and the atmosphere, oceans and the subterraneous riches, the whole orb which is the seat of our life and experience, heavenly objects both old and new, and the elements out of which these things seem to be composed: earth, water, air, and fire.".

While the science of mechanics grew out of applications of mathematics to understand physical phenomena, a relatively new, more universally applicable approach, is the combination of mathematics and high performance computers. This newer approach is call *computational mechanics* or *computational engineering science*. This approach includes the capability to study physical phenomena over multiple scales (multiscales) that can predict phenomena at one scale based upon knowledge of phenomena at smaller scales. Scales may range from that of a single atom to large-scale systems such as airplanes, oceans, and climate systems.

A recent book that covers a wide spectrum of mechanics is Tinsley Oden's *An Introduction to Mathematical Modeling: A Course in Mechanics* (Oden 2011). Oden wrote his book for an introductory class in the University of Texas program Computational Science, Engineering, and Mathematics. The book covers Nonlinear Continuum Mechanics, Electromagnetic Field Theory and Quantum Mechanics, and Statistical Mechanics, all from a modern point of view.

The present book is concerned primarily with the people of classical mechanics, as opposed to relativistic and quantum mechanics that are more the purview of modern physics. This book presents a history of the U.S. National Committee on Theoretical and Applied Mechanics and its relationship with the International Union of Theoretical and Applied Mechanics (IUTAM). The USNC/TAM represents the mechanics community in the United States. The members of the US National Committee on Theoretical and Applied Mechanics are mostly engineers and applied mathematicians who, typically, work in classical mechanics. After 18 years functioning as an independent committee, the USNC/TAM agreed to become a committee under the umbrella of the National Academies in 1966.

The following broad overview will give the reader an appreciation for the people, their backgrounds and activities, a time history of major events, centers of major activities, and advancements in the engineering science of mechanics.

Brief History of Mechanics

Mechanics has a long and rich history as an engineering science. The earliest recorded activity was the works of Aristotle and Archimedes in classical antiquity. Classical antiquity refers to the long (eighth century BC to fifth century AD) period of cultural history of the Greco-Roman world that centered on the Mediterranean Sea. There was very little advancement in the field after Archimedes for nearly 1800 years. The pace of activity increased in the sixteenth century with the work of Leonardo da Vinci and Galileo, and accelerated after the work of Newton in the seventeenth century. The pace of activity continued to increase through the eighteenth and nineteenth centuries, followed by an exponential growth during the twentieth century. The emphasis on the use of applied mathematics for mechanics at the University of Göttingen, Einstein's relativity theory, and Planck's quantum mechanics all appeared early in the twentieth century.

The map below shows the cradle for the growth of what might be termed "*modern mechanics*" (Fig. 1.1). It ranged from Cambridge, England in the west (Newton) to Kiev (Timoshenko) in the east. Göttingen (Prandtl and von Kármán) is near the center of the cradle. Aachen, Paris, Delft, St. Petersburg and Budapest also provided significant contributions to the growth of the science. The following pages give brief descriptions of many of the individuals who made important contributions to the field.

Fig. 1.1 The cradle of mechanics

The early history of mechanics suffers, at times, from a lack of verifiable documentation. Often, the individual who received credit for a discovery or development built upon previous developments. Two books, *History of Mechanics* (Dugas 1955) and *A Concise History of Mathematics* (Struik 1987), proved particularly helpful when studying the history of mechanics. The fact that Struik's book on mathematics was so helpful demonstrates the close association between mathematics and mechanics.

The history of mechanics is discussed in numerous books and journal articles, and, in today's world, on the web.

Writings that include substantial historical information are provided in a separate listing under the heading References with History.

Chapter 2
The Giants of Mechanics

In the following, I present brief descriptions of individuals who have made major contributions to the field of mechanics. For each individual, their scholarly activities follow a listing of their place of birth, primary associations, and lifespan. The public is aware of many of these individuals, at least by name; those who have experience in science and engineering will recognize most of the others. Without doubt, the listing is somewhat selective, and others might have been included as well. The more I researched the topic, the more scholars I wanted to include in the listing.

Ancient Greece (BC)

Aristotle

Aristotle (Stagirus, Macedonian (now Greece), 384–322 BC) an ancient Greek philosopher and scientist was a student of Plato and a teacher to Alexander the Great. His writings covered a wide variety of subjects including physics, biology, zoology, metaphysics, logic, ethics, aesthetics, poetry, theater, music, rhetoric, linguistics, politics, and government. Treatises related to mechanics are the eight books of *Physics* (Aristotle, 350 BC, see Hardie and Gaye 1930, translation) and *On the Heavens* (Aristotle, 350 BC, translation by J. L. Stocks 1922). He discusses the "law of the lever" in his treatise "Mechanical Problems". Aristotle's writings are philosophical and formed the general principles for understanding the nature of moving things for nearly 2000 years. Followers of his works were called Peripatetics, after the Lyceum in Athens where members met. Classical mechanics of the sixteenth century disproved many of his basic ideas.

© Springer International Publishing Switzerland 2016
C.T. Herakovich, *Mechanics IUTAM USNC/TAM*,
DOI 10.1007/978-3-319-32312-1_2

Archimedes

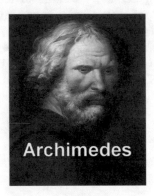

Archimedes (Syracuse, Sicily, 287–212 BC) an Ancient Greek mathematician contributed to mathematics, astronomy, statics, and hydrostatics, i.e., *Archimedes principrle*. Many writers and scholars consider Archimedes to be the greatest mathematician of antiquity (see for example Struik, *A Concise history of Mathematics*, page 50). He was one of the first to apply mathematics to physical phenomena, founding hydrostatics and statics, including an explanation of the principle of the lever and buoyancy. His credits include designing innovative machines, such as a screw pump, compound pulleys, and defensive war machines to protect his native Syracuse from invasion. His most important contributions in mathematics may have been in what we now call *integral calculus*.

Little is known about the personal life of Archimedes.

The Renaissance (Roughly 1350–1700)

Leonardo da Vinci

Leonardo di ser Piero da Vinci (Vinci, Florence, Italy, 1452–1519) is credited with being a multifaceted, Renaissance man as painter, sculptor, architect, mathematician, engineer, and inventor, among other things. Leonardo was born out of wedlock to a notary, Piero da Vinci, and a peasant woman, Caterina; he was self-taught. He is most famous for his paintings *Mona Lisa*, *The Last Supper*, *The Creation of Adam* on the ceiling of the Sistine Chapel, and his drawing of the *Vitruvian Man*. There is considerable debate as to the originality of his many mechanical drawings, but there is no question that he wrote extensively on mechanics. Truesdell devoted 80 pages to da Vinci in his *Essays in the History of Mechanics* (1968). Regarding da Vinci's fame he writes, "First, there is the enormous *bulk* of what Leonardo wrote on mechanics, scientific or unscientific, right or wrong. No one before him had ever written *so much* on mechanics, and few afterward have done so".

It is noteworthy that the interval between the work of the ancient Greeks (Aristotle and Archimedes) and da Vinci was nearly 2000 years.

Galileo Galilei

Galileo Galilei (Pisa, Italy, University of Pisa, University of Padua, 1564–1642) known simply as Galileo, was a physicist, mathematician, engineer, astronomer, and philosopher. Galileo's, *Dialogues Concerning Two New Sciences* (1638) summarizes his work over the preceding 30 years on kinematics and strength of materials. His main contributions were the development of improved telescopes, astronomy, showing that Aristotle was wrong about relationship between the speed of a falling object and its weight, and writing that the Earth and solar system rotated about the sun as in Copernican theory. For the latter, in 1634, the Catholic Church found him guilty of heresy; he was confined to house arrest for the remainder of his life. He died in Arcetri, Italy on January 8, 1642.

Robert Hooke

Robert Hooke (Freshwater, Isle of Wight, England; Oxford, 1635–1703) was an English natural philosopher who proposed the linear elastic response of solids, now known as Hooke's Law. His law continues to be the fundamental basis for elasticity theory. Hooke was good at building and operating devices, including the vacuum pump used in Robert Boyle's gas law experiments, telescopes, and microscopes. He had an excellent facility with experiments.

Hooke had a contentious dispute with Newton over who was the first to state the law of gravitation. He also had a dispute with Huygens over who invented the balance spring that enabled having a watch as a timepiece. Hooke is generally given the nod as the first to invent the balance spring, but Newton is generally considered to be the first with the law of gravitation.

Isaac Newton

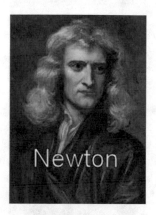

Sir Isaac Newton (Woolsthorpe-by-Colsterworth, Lincolnshire, England; Wadham College, Oxford, Trinity College, Cambridge, 1642–1727) was an English physicists and mathematician who is famous for his three laws of motion, the law of gravitation, and the development of calculus. Newton laid the foundation for *classical mechanics* (also called *Newtonian Mechanics*) that continues to be a fundamental building block for mechanics in the twenty-first century. The exceptions being when Einstein's relativity theory and Planck's quantum mechanics are required. Newton's *Philosophiae Naturalis Principia Mathematica* ("*Mathematical Principles of Natural Philosophy*") was first published in 1687. In *Principia*, Newton presented his three laws of motion and the law of gravitation.

Newton was, at times, modest of his own achievements, famously writing in a letter to Robert Hooke in February 1676:

"If I have seen further it is by standing on the shoulders of giants."

Nevertheless, Newton strongly defended his belief that he, rather that Hooke, was the first to propose the law of gravitation, and his belief that he, rather than Leibniz, developed calculus. According to Struik (page 106), both men did their work developing calculus independently, with Newton doing his work first (1655–1666), and Leibniz publishing his work (1684–1686).

Newton's date of birth was either December 25, 1642, or January 4, 1643 depending upon which calendar is used. The 1643 date is consisted with the new (current) Gregorian calendar. He died in 1727 at age 85.

Industrial Revolution (Roughly 1700–1850)

Daniel Bernoulli

Daniel Bernoulli (Groningen, the Netherlands; Imperial Russian Academy of Sciences, St. Petersburg, University of Basel, 1700–1782) was from a distinguished family of mathematicians. His family moved to Basel, Switzerland when he was quite young and he was thus considered to be a Swiss mathematician. He is noted for his work in fluid mechanics, and, in particular, the conservation of energy identified simply as *Bernoulli's equation*. During an early career as a medical doctor he studied the mechanics of breathing, flow of blood and blood pressure. He was a colleague of Euler for 6 years in St. Petersburg. Together they did work on vibrating systems and developed Euler-Bernoulli beam theory. Bernoulli also wrote *Hydrodynamica* (1738) while in St. Petersburg, and thereby coined the term *hydrodynamics*. He won the Grand Prize of the Paris Academy ten times writing on a wide range of topics.

Leonard Euler

Leonhard Euler (Basel, Switzerland; University of Basel, Imperial Russian Academy of Sciences, St. Petersburg, Berlin Academy, 1707–1783) a Swiss mathematician did most of his work in St. Petersburg and Berlin. He was a pupil of Johann Bernoulli, Daniel's father, and colleague of Daniel Bernoulli. He was married twice and had 13 children. He was blind by age 59, but continued to be very productive (with the aid of an assistant) as he had a phenomenal memory.

His is known for his book *Mechanica* (1736) in which he introduced analytical methods for Newtonian mechanics, the Euler-Bernoulli beam equation, calculus of variations, the Euler buckling load for column instability, and the Euler-Lagrange equation for optimization of a functional. Euler introduced numerous mathematical notations that are in common use today: the concept of functions $f(x)$, trigonometric function notation, the letter e for the base of the natural logarithm (Euler's number), Σ for summations, and i for imaginary unit.

Struik refers to Euler as the "most productive mathematician of the eighteenth century—if not all time."

Paris and École Polytechnique (Eighteenth and Nineteenth Centuries)

Jean-Baptiste le Rond d'Alembert

Jean-Baptiste le Rond d'Alembert (Paris, France, 1717–1783) was a self-taught French mathematician who made contributions in dynamics and fluid mechanics. He was the illegitimate child of the writer Claudine Guérin de Tencin and the chevalier Louis-Camus Destouches, an artillery officer. His mother left him on the steps of the church Saint-Jean-le-Rond de Paris a few days after he was born. According to custom, he was named after the patron saint of the church.

D'Alembert's *Traité de Dynamique* was first published in 1743 and updated in 1758. His is known for d'Alembert's principle in dynamics, his work on the vibrating string and the development (with Daniel Bernoulli) of partial differential equations, d'Alembert's paradox on drag in fluid flow, and the one-dimensional wave equation known as d'Alembert's equation.

Joseph-Louis Lagrange

Joseph-Louis Lagrange (Turin, Piedmont-Sardinia, Italy; Prussian Academy of Sciences, Berlin, École Polytechnique, Paris, 1736–1813) was an Italian–French mathematician who made major contributions in mathematics and mechanics. His major contributions include calculus of variations, Lagrangian mechanics, Lagrange multipliers, Euler-Lagrange equations, and his comprehensive treatise *Mécanique analytique* (Lagrange, 1788). In this book, Lagrange applied analytical methods to present a unified treatment of the mechanics of points and rigid bodies.

Although born in Italy, Lagrange worked in Berlin for 20 years before completing his career in Paris. He was invited to Berlin in 1766 by Frederick the Great when Euler left Berlin for St. Petersburg. Lagrange was the first professor of mechanics at École Polytechnique when it opened in Paris in 1794. His name is one of 72 inscribed on the Eiffel Tower.

Pierre-Simon Laplace

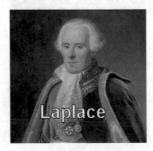

Pierre-Simon Laplace (Beaumont-en-Auge, Normandy, France; University of Caen, École Militaire, 1749–1827) was a French mathematician who worked in celestial mechanics and statistics. He became professor of mathematics at École Militaire (the military school) in Paris with assistance from D'Alembert. A friend of Napoleon, he also took part in the organization of École Normale and École Polytechnique. His work in celestial mechanics (Laplace, *Mécanique céleste*, 5 *Vols.*, *1799–1825*) was the first such work based on calculus. He formulated Laplace's equation and developed the Laplace transform, two mathematical ideas used extensively in the study of mechanics. His name is one of 72 inscribed on the Eiffel Tower.

Claude-Louis Navier

Claude-Louis Navier (Dijon, France; École Polytechnique, École Nationale des Ponts et Chaussées, 1785–1836) was a French engineer and physicist who made significant contributions in both solid mechanics and fluid mechanics. He is known (with Stokes) for the Navier-Stokes equation in fluid mechanics. In solid mechanics, he formulated the equations for the general theory of elasticity in a mathematically usable form (paper read to the Académie des Sciences, Paris, 1821).

Navier studied at École Polytechnique and École des Ponts et Chaussees. In addition to his fundamental studies, Navier participated in much practical engineering work on the design and construction of bridges in particular. Like other leading mathematicians of that era, he taught at École Polytechnique. His name is one of 72 inscribed on the Eiffel Tower.

Augustin-Louis Cauchy

Augustin-Louis Cauchy (Paris, France; École Nationale des Ponts et Chaussées, École Polytechnique, 1789–1857) was a French mathematician who applied his expertise in applied mathematics and elasticity. As a young man his family lived in Arcueil, near Paris, where he became acquainted with other mathematicians including Laplace and Lagrange. Cauchy introduced the concept of stress and strain components, and the 3×3 symmetric matrix known at the Cauchy stress tensor (Cauchy 1827). He showed that, for an anisotropic material, at most 15 constants are required to define the elastic properties and that only two elastic constants are required for an isotropic material. He also introduced the concept of principal directions and principal stresses. Cauchy was the first to demonstrate that non-circular bars warp when subjected to torsional loading.

Cauchy defined complex numbers as pairs of real numbers and provided a definition for a continuous function $f(x)$. With Reimann, he is given credit for the Cauchy-Reimann equations. His name is one of 72 inscribed on the Eiffel Tower.

Siméon Deni Poisson

Siméon Deni Poisson (Pithiviers, Loiret, France; École Polytechnique, 1781–1840) was a French mathematician and physicist. He contributed in celestial mechanics, rational mechanics, electricity, and magnetism. Poisson's ratio is a fundamental property for material response, and Poisson's equation is well-known to those working in mechanics and electrostatics. Poisson obtained the three equations of equilibrium and boundary conditions working from a molecular point of view. He also showed that if a disturbance is produced in a small portion of a body, it results in two kinds of waves, dilatational and distortional. His *Treatise de mecanique* (1811 and 1833) is a classic covering a broad spectrum of topics in mechanics.

Poisson had close personal and professional associations with both Lagrange and Laplace. Upon graduation from École Polytechnique he stayed on as instructor of mathematics, eventually becoming professor. His name is one of 72 inscribed on the Eiffel Tower.

George Green

George Green (Sneinton, Nottinghamshire, England; Cambridge, 1793–1841) was a British mathematician who had no more than 1 year of early formal education. He was self-taught and published several very important works prior to entering Cambridge University at age 40. In 1828, at age 35 he published *An Essay on the Application of Mathematical Analysis to the Theories of Electricity and Magnetism*. This paper had several important mathematical concepts that are in use today, *Green's theorem*, *Green's functions*, and the *concept of potential functions*. His 1837 (and 1839) paper proposed the concept of strain-energy density. This concept led to the accepted theory that an anisotropic material has at most 21 independent material constants. This settled a longstanding argument as to whether there were 15 or 21 independent constants. Green completed his degree at Cambridge at age 46, only to die 2 years later at age 48.

Green never married but had seven children with Jane Smith, daughter of William Smith, manager of his father's corn-mill.

Adhémar Jean Claude Barré de Saint Venant

Adhémar Jean Claude Barré de Saint Venant (Villiers-en-Bière, Seine-et-Marne, France; École Polytechnique, École des Ponts et Chaussées, 1797–1886) was a mechanician and mathematician who made important contributions in solid mechanics, elasticity, torsion, viscosity, derivation of the Navier-Stokes equations and hydrodynamics. *Saint Venant's principle* for the equivalence of loadings at a distance from the points of application may be his most well-known contribution.

Saint Venant encountered difficulty with other students at Ecole Polytechnique during the military actions of 1814. He was proclaimed a deserter and not allowed to complete his studies at Ecole Polytechnique. Finally, in 1823, he was allowed to enter the Ecole des Ponts et Chaussees where he was shunned by other students, but still graduated first in his class. In 1837 he was asked to give lectures at the Ecole des Ponts et Chaussees when Professor Coriolis became ill. These lectures were lithographed and are the most complete written record of his work.

Technological Revolution (Roughly 1850–1910)

William Rowan Hamilton

William Rowan Hamilton (Dublin, Ireland; Trinity College, Dublin 1805–1865) was an Irish mathematician who worked in classical mechanics, optics, and mathematics. He attended Trinity College, Dublin, and remained there as a professor for the remainder of his career. He reformulated *Newtonian mechanics* as a variational problem called *Hamiltonian mechanics*. Hamilton's variational *principle* is applicable for classical mechanics, electromagnetic and gravitational fields; and it has been extended for quantum mechanics. His name is also associated with the *Hamilton-Jocobi equation* in the calculus of variations.

Hamilton is considered to be Ireland's most famous scientist. He was the first Foreign Associate elected to the U.S. National Academy of Sciences in 1864.

George Gabriel Stokes

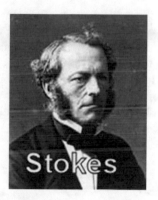

George Gabriel Stokes (Skreen, County Sligo, Ireland; Bristol College, Pembroke College, Cambridge University, 1819–1903) was an Irish mathematician who moved to Bristol, England at age 16 to study at Bristol College. He then studied at Pembroke College at Cambridge where he stayed following completion of his studies, eventually being appointed as Lucasian Professor of Mathematics. His major contributions are the Navier-Stokes equation in fluid mechanics, and Stokes theorem in vector calculus relating a surface integral to a line integral over its boundary.

Hermann Ludwig Ferdinand von Helmholtz

Hermann Ludwig Ferdinand von Helmholtz (Potsdam, Germany, Universities of Königsberg, Bonn, Heidelberg and Berlin, 1821–1894) was a German medical doctor as well as a physicist. Much of his work was in physiology; it would fall into the realm of biomechanics using current terminology. His most significant work was the discovery of the principle of conservation of energy while studying muscle metabolism. He also developed the Helmholtz equation for problems involving partial differential equations in both space and time. Helmholtz also studied the hydrodynamics of vortex motion, optics, and acoustics.

Gustav Robert Kirchhoff

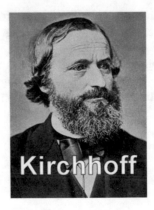

Gustav Robert Kirchhoff (Königsberg, Russia; Universities of Königsberg, Berlin, Breslau and Heidelberg, 1824–1887) was a German physicist who made contributions on a range of subjects including electric circuits, spectroscopy, black-body radiation, and mechanics. In mechanics, the Kirchhoff-Love plate theory and the Piola-Kirchhoff stress tensor are two significant contributions.

At Konigsberg, Kirchhoff was a student of Franz Neumann and attended a seminar directed by Jacobi, Neumann, and Richelot. Kirchhoff married Richelot's daughter. At Breslau, he met and worked with Bunsen and moved to Heidelberg with Bunsen. There they were joined by Helmholtz to form the basis of an outstanding scientific era at Heidelberg University.

The Twentieth Century

The pace of activity in the science of mechanics accelerated early in the twentieth century. Strong activity in England, and Europe, in particular Germany, led the developments. Some of the researchers who were active at that time are presented in the following.

Lord Rayleigh

Lord Rayleigh (*John William Strutt*), (Maldon, United Kingdom; Trinity College, Cambridge, 1842–1919) was an English physicist who was recognized with the 1904 Nobel Prize for Physics for discovering Argon. He contributed in mathematics, wave propagation, acoustics, light, elasticity and hydrodynamics. *Rayleigh waves* are the waves on the surface of a homogeneous, isotropic, semi-infinite solid. *Rayleigh's equation* refers to beam vibration including rotary inertia effects. The Rayleigh-Ritz method provides approximate solutions of differential equations. Rayleigh's *Theory of sound* (Vol. 1, 1877 and Vol. 2, 1878) are classics.

Osborne Reynolds

Osborne Reynolds (Belfast, Northern Ireland; Queens' College, Cambridge, Owens College, Manchester, 1842–1912) was born in Belfast but moved to Dedham, England shortly thereafter. He completed his education at Queens College, Cambridge; 1 year later, he was appointed *Professor of Engineering* at Owens College, Manchester. His primary interest was fluid mechanics. His experiments on fluid flow in pipes lead to the definition of *Reynolds number* for defining flow as either laminar or turbulent. He also studied heat transfer at the interface between fluids and solids.

Horace Lamb

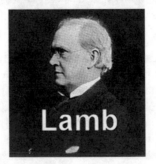

Horace Lamb (Stockport, England, Owens College, Manchester, Trinity College, Cambridge, University of Adelaide, Victoria University of Manchester, 1849–1934) was an English applied mathematician who made major contributions in hydrodynamics, wave motion, and theory of sound. At Trinity College his professors included Stokes and Maxwell.

His Treatise on the Mathematical Theory of the Motion of Fluids (name later changed to Hydrodynamics, Lamb, 1932) was published in six editions between

1879 and 1932. Between 1895 and 1928, he wrote books on Statics, Dynamics, Infinitesimal Calculus, Elastic Waves, Dynamics of Sound, Higher Mechanics, and Mathematical Physics. The first American edition of Hydrodynamics, published in 1945 by Dover, continues to be a classic. His work on wave propagation in thin layers is recognized by their name, Lamb waves.

Christian Felix Klein

Christian Felix Klein (Düsseldorf, Germany; Universities: Bonn, Munich's Technische Hochschule, Leipzig, Göttingen; 1849–1925) was a German mathematician who emphasized the application of mathematics to real-world problem. His mathematical contributions were in group theory. His most significant contributions may have been in organization and leadership. He is credited with the establishment of the Göttingen Institute for Aeronautical and Hydrodynamical Research. This Institute had a major impact on the growth of mechanics as an engineering science.

A number of mechanicians of his era credit Klein with instilling in them the values of applied mathematics. More will follow on this point in a later discussion of the impact of Göttingen on mechanics.

Ludwig Prandtl

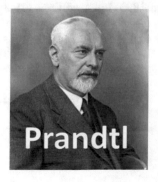

Ludwig Prandtl (Freising, Germany; Technical University Munich, Hannover, Göttingen, 1875–1953) was the son of a German professor of engineering. At the Technical University Munich Prandtl's Ph.D. dissertation was on elastic stability under the direction of his advisor August Foeppl. For several years after his Ph.D., he worked in industry where he became interested in fluid mechanics. In 1901 he was appointed professor of fluid mechanics at what is now the Technical University of Hannover. At Hannover he did much of his groundbreaking work on boundary layers, drag, and flow separation.

As a result of his outstanding work at Hannover, he was hired at Göttingen as professor of applied mechanics and director of the Institute for Technical Physics. Prandtl developed this institute into a world-class aerodynamics research and educational organization that attracted students from around the world. Two students who worked with him were Theodore von Kármán and Stephen Timoshenko, both of whom worked in solids mechanics.

The *Prandtl number* in boundary layer heat transfer carries his name as does the crater *Prandtl* on the far side of the Moon. Many consider Prandtl to be the father of modern aerodynamics.

Augustus Edward Hough Love

Augustus Edward Hough Love (Weston-super-Mare, England; St. John's College, Cambridge, Oxford, 1863–1940), more commonly known as A. E. H. Love, was a mathematician famous for his work on the mathematical theory of elasticity and wave propagation. His *A Treatise on the Mathematical Theory of Elasticity* appeared in two volumes in 1892 and 1893. Later editions (1906, 1920 and 1927) incorporated significant improvements. For many years, Love's book was considered the standard mathematical work on elasticity. In 1911, he developed a mathematical model for surface waves now known as *Love waves*. This work was published in 1911 as a book entitled *Some Problems of Geodynamics*.

Albert Einstein

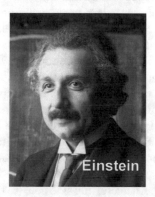

Albert Einstein (Ulm, Württemberg, Germany; Universities of Bern, Prague, Zurich, Berlin, Caltech and Princeton 1879–1955) was a German-born physicist. His father was a salesman and engineer. Einstein published four papers in 1905 (Einstein 1905a, b, c, d) that were to have a major impact on our understanding of the physical world. In particular, his *Special Theory of Relativity* showed that there are limits on the validity of *classical (Newtonian) mechanics*; his work on photons opened the door to the development of *quantum mechanics*. He received the 1921 Nobel Prize in Physics *"for his services to theoretical physics, and especially for his discovery of the law of the photoelectric effect"*. His most well-known development, by the general public, is the equation $E = mc^2$ for the equivalence of energy and mass.

Einstein was a celebrated worldwide lecturer. He was friends with a wide variety of people ranging from film star Charlie Chaplin to physicist Max Born. He was visiting in the United States in 1933 when Adolf Hitler was elected chancellor. He never returned to Germany and became an American citizen in 1940. During the persecution of Jews in the 1930s, he used his influence to encourage those in positions of power to assist Jewish scientists to leave Germany. He died at age 76 in Princeton, NJ where he was a resident scholar in the Princeton Institute for Advanced Study.

Richard V. Southwell

Richard V. Southwell (Norwich, England; Trinity College, Cambridge, Oxford, Imperial College, London, 1888–1970) was a British mathematician who applied his mathematical expertise for approximate solutions of problems in applied mechanics. Following service in World War I he was head of the Aerodynamics and Structures Division at the Royal Aircraft Establishment, Farnborough. He then moved, in order, to the National Physical Laboratory, Trinity College, and Oxford University. During this time, he developed relaxation methods for solving partial differential equations that were used extensively for problems in solid and fluid mechanics. He completed his career as Rector at Imperial College, London. One of his students, Olgierd Zienkiewicz, was a pioneer of the finite element method.

Southwell was one of those involved early on in the establishment of the International Union of Theoretical and Applied Mechanics (IUTAM). ASME recognized his contributions with both the Worcester Reed Medal in 1941 and Timoshenko Medal in 1959. An interesting side note (to this author) is that Southwell and G. I. Taylor often played golf together on Sundays and during weekend outings with family and friends.

Geoffrey I. Taylor

Geoffrey I. Taylor (St. John's Wood, London; Trinity College, Cambridge; 1886–1975) was a British mathematician and physicist. His major contributions were in fluid mechanics and wave propagation; he also made contributions in solid mechanics. His early work was on shock waves, and turbulence in the atmosphere and in oceanography. His work in solid mechanics included research on the deformation of crystalline materials and plastic deformation of ductile materials in terms of dislocations.

During World War II he assisted on the Manhattan Project at Los Alamos in the United States. Taylor was one of those involved early on in the establishment of the International Union of Theoretical and Applied Mechanics (IUTAM). ASME recognized Taylor with the Timoshenko Medal (1958). ASCE recognized him with the Theodore von Kármán Medal (1969), and he received the Franklin Medal in 1962.

Johannes (Jan) M. Burgers

Johannes (Jan) M. Burgers (Arnhem, the Netherlands; Leiden, Delft and Maryland; 1895–1981) was a Dutch physicist who studied under Paul Ehrenfest. He made fundamental contributions in dislocation theory and viscoelasticity. At Leiden he became acquainted with Albert Einstein and Niels Bohr; a fellow student was D. J. Struik.

Burgers' early work was on fluid dynamics at low Reynolds number and was related to Prandtl's work on airfoils. He did important work on turbulence using hot-wire anemometry. This work led to what is now known as *Burgers' equation* for one-dimensional, nonlinear differential equations. He introduced *Burgers' vector* when working on dislocation in crystal lattices (with his brother).

Burgers is considered a co-founder of IUTAM having developed a close relationship with von Kármán at the 1922 Innsbruck conference. Eventually he wrote the formal plan for the establishment of IUTAM in 1946.

More on Burgers is presented in the later section of this book on immigrants to the United States.

Tullio Levi-Civita

Tullio Levi-Civita (Padua, Italy; University of Padua, University of Rome, 1873–1941) was an Italian mathematician. He is recognized for his work in tensor calculus, celestial mechanics, and hydrodynamics. At Padua he was appointed Chair of Rational Mechanics and at Rome he was appointed Chair of Mechanics.

Einstein relied on Levi-Civita's tensor calculus theory when developing his theory of general relatively. As a result of this working relationship, they developed a close friendship. Levi-Civita actually lived in Einstein's home at Princeton for a year in 1936. When war seemed imminent, he returned to Italy; however, he was soon deprived of his professorship and membership in all scientific societies because he was a Jew.

Levi-Civita was the first person that von Kármán asked to assist with the organization of the 1922 Innsbruck Conference on Applied Mechanics. Einstein once remarked that what he liked best about Italy was "spaghetti and Levi-Civita".

Stephen P. Timoshenko

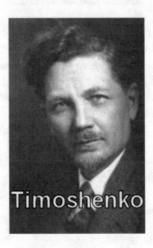

Stephen Prokof'yevich Timoshenko, (Shpotivka, Ukraine; St. Petersburg Polytechnic, Michigan and Stanford; 1878–1972), was a Russian (Ukraine) applied mathematician whose most significant contributions were his books in solid mechanics. Timoshenko was born on December 23, 1878 in Shpotovka in the Ukraine, which was part of the Russian Empire at the time. He completed his secondary education with a gold medal in Romny, near Kiev. He then studied engineering at the Institute of Ways of Communication in St. Petersburg, graduating in 1901.

Prior to his graduation, he had made two trips to Western Europe. As will be shown, he was to make many more trips to Western Europe throughout his life. Following graduation and travels, he returned to St. Petersburg to work in the Mechanics Laboratory of the Ways of Communication Institute testing materials and assisting in lectures on mathematics.

In 1903, Timoshenko became an instructor at St. Petersburg Polytechnic Institute. He spent the summer of 1904 in Western Europe including a brief stop at Berlin Polytechnic, 6 weeks with Professor A. Föppl in Munich, and then went to Paris for Bastille Day. While in Paris, he visited the Louvre and the *Arts et Métiers* laboratory. By late July, he (and his wife) were at Lake of Thun in Switzerland where he worked on a translation of Love's Theory of Elasticity. They returned to St. Petersburg via Zurich and Vienna.

His travels in Europe during the summer of 1904 are typical of how Timoshenko spent many summers throughout his life. The Lake of Thun was a favorite place. He and his wife would visit there often during their summers in Europe. Much of this time was spent working (alone) on his writings.

The Polytechnic Institute at St. Petersburg was closed during the 1904–1905 academic year as a result of student demonstrations and general unrest due to the war with Japan. Timoshenko went to Göttingen, Germany in April 1905 to work on stability under the guidance of Professor Ludwig Prandtl. He returned to Göttingen in the summer of 1906 for additional studies before completing his doctoral degree at Kiev Polytechnic in 1907; Prandtl was listed as Advisor Number 2 on his dissertation. Timoshenko returned to Göttingen for more studies from April to September, 1909.

During this period of the early twentieth century, Timoshenko became convinced that engineers needed a stronger background in the sciences and mathematics. The works of the Russian physicist Aleksey N. Krylov, the German mathematician Felix Klein at Göttingen, and the mechanical engineer A. Stodola in Zurich were three people who led him to this conclusion.

Timoshenko held various teaching positions in Russia, and a year of military service, during the period 1906–1920. This was often a time of political unrest and difficulty for him and his family. As a result of the unstable situation in Russia, he decided to go to Yugoslavia which had indicated a willingness to accept Russian refugees. With difficulty, he managed to relocate himself and his family to Zagreb where he completed two productive years, 1920–1922, at Zagreb Polytechnic.

More on Timoshenko is presented in the later section of this book on immigrants to the United States.

Paul Ehrenfest

Paul Ehrenfest (Vienna, Austria; Vienna, Göttingen, Leiden, 1880–1933) was an Austrian who worked in theoretical physics. His most significant contributions were in statistical mechanics and quantum mechanics. He earned his Ph.D. at Vienna where he took courses from Ludwig Boltzmann. During his graduate studies, he studied for a time at Göttingen, and returned there following completion of his degree. Following Boltzmann's suicide, Felix Klein at Göttingen invited Ehrenfest to write a review on statistical mechanics that Boltzmann had agreed to write.

Following time in St. Petersburg, Ehrenfest settled at Leiden, the Netherlands in 1912 where he stayed for the remainder of his career. He was a close friend of Albert Einstein and Niels Bohr. Einstein would stay at Ehrenfest's home when he visited Leiden—as he often did. Ehrenfest had many students including Johannes Burgers. Unfortunately, depression overcame Ehrenfest in 1933 and he first shot his son who had Down syndrome and then shot himself.

Theodore von Kármán

Theodore von Kármán (Budapest, Hungary; Göttingen, Aachen, Caltech; 1881–1963) was a Hungarian applied mathematician who worked in both fluid mechanics and solid mechanics. He was born on May 11, 1881, in Budapest, Austria–Hungary (at the time). His full name in Hungarian was von Sköllöskislaki Kármán Todor. In English, he was known as Theodore von Kármán, however, family and friends often referred to him as Todor. He studied at what is now the Budapest University of Technology and Economics, receiving his undergraduate degree in mechanical engineering in 1902. Following a year of military service, he returned to the Technical University in a teaching capacity for 3 years.

He then went to Germany to study under Ludwig Prandtl at the University of Göttingen. He arrived at Göttingen in October, 1906. Prandtl was only 33 years old (8 years older than von Kármán), and Head of the Department of Applied Physics when von Kármán arrived in Göttingen. Prandtl had already gained prominence as a scientist, primarily because of his pioneering work in fluid mechanics, but he had also done work in other areas of applied mechanics. von Kármán told Prandtl that he wanted to work on stability. He received his doctorate, under Prandtl's guidance, in 1908. His dissertation was concerned with inelastic buckling of columns.

Prandtl had been hired by Felix Klein, the promoter of Göttingen's research activity in applied mathematics and applied mechanics. At Göttingen, Felix Klein and David Hilbert impressed von Kármán with the importance of mathematics. von Kármán attributed the origin of modern applied mechanics to Klein. Klein believed very strongly, as did von Kármán, in the need for fundamental mathematics underpinning mechanics principles. A fellow student, colleague, and friend at Göttingen was Max Born who later won the Nobel Prize for his contributions to quantum physics.

Following completion of his doctorate, von Kármán accepted a position as *privat dozent* (a candidate for the faculty) at Göttingen to work with Prandtl assisting in building a wind tunnel, conducting research in fluid mechanics and teaching mechanics courses. He stayed at Göttingen in this capacity for 4 years. During this time he published the work that he is most well-known for, the Kármán vortex street.

In 1912, unhappy that he had not received a permanent faculty appointment at Göttingen, von Kármán accepted a position as Director of a new Aeronautical Institute at the Technical University in Aachen, Germany. He remained at Aachen (with the exception of 4 years back in Hungary during the First World War), until 1930. While at Aachen, von Kármán earned an international reputation for his contributions to aerodynamics. He built the new institute into one with worldwide acclaim, attracting students from many countries.

The atmosphere among the faculty and students at Aachen was much more relaxed than it was at Göttingen, largely due to von Kármán's personality. He was a gregarious individual who liked to interact with people and have parties. He also liked to tell jokes, often of the risque variety. During his years at Aachen, he lived a short distance across the border in Vaal, Holland, with his mother and sister. He would take the trolley (approximately ten miles) to work each day. The von Kármán household was large and they had many parties with students, assistants, and visitors in attendance. Mother and sister were heavily involved making sure that there was plenty to eat and drink at these get-togethers that took place during many weekends. Many of the attendees were French, Italian, Hungarian, Dutch, English, and German. von Kármán's mastery of many languages was a great asset, for he spoke fluent Hungarian, German, French, Italian, Yiddish, and what he always described as the international language, "bad English" (see Dryden, NAS memoir on von Kármán).

While much of von Kármán's early work was in solid mechanics (stability in particular) and physics (he and physicist Max Born developed the concept of the crystal lattice), he is more widely recognized for his contributions in fluid mechanics. In his autobiography (page 59) von Kármán makes the statement "... aerodynamics, the science of flight, which became my major interest in life". The Kármán vortex street (or sheet) (published in 1911 and 1912) is named for him. Vortex shedding is an important consideration in many applications of flow past solid objects. He did this initial work while at Göttingen. In 1940, he used the analysis of the vortex street to explain the collapse of the Tacoma-Narrows Bridge in the state of Washington. The existing wind conditions caused unstable vortex shedding from the bridge, resulting in severe oscillations and eventual collapse. He also used vortex shedding to explain other, previously unexplained, phenomena.

Not realizing at the time (1962) that my master's thesis work was related to von Kármán, your author identified unstable vortex shedding at flow from a main pipe into symmetrical lateral pipes as the cause of significant vibration (bouncing) of pipes at a petroleum barge loading dock. This work included building a wind tunnel and a smoke generator in order to visualize the airflow and thus the vortices. I wish I had known about von Kármán's work—no Google then!

More on von Kármán is presented in the later section of this book on immigrants to the United States.

Richard Edler von Mises

Richard Edler von Mises (Lemberg, Austria (now Lviv, Ukraine); Vienna, Brünn, Strasburg, Dresden, Berlin, Istanbul, Harvard; 1883–1953) was a mathematician who worked in applied mathematics, solid mechanics, fluid mechanics, and aerodynamics. Mises was born in Lemberg, then part of Austria–Hungary. He attended the Akademisches Gymnasium in Vienna, from which he graduated with honors in Latin and mathematics in 1901. He then majored in mathematics, physics, and engineering at the Vienna University of Technology. In 1905, still a student, he published an article in the prestigious Zeitschrift für Mathematik und Physik.

In 1908 Mises completed his doctorate from Vienna, and received his habilitation from Brno (Czech Republic) to lecture on engineering. In 1909, at age 26, he was appointed professor of applied mathematics in Straßburg, then part of the German Empire (now Strasbourg, Alsace, France). He had learned to fly and gave the first university course on powered flight in 1913. With the outbreak of the First World War, he joined the Austro-Hungarian army where he served as a test pilot, instructor, and supervisor for the "Mises-Flugzeug" airplane.

Following the war, von Mises was appointed as the new chair of hydrodynamics and aerodynamics at the Dresden Technische Hochschule. In 1919, he was appointed Director of the new Institute of Applied Mathematics at the University of Berlin. In 1921 von Mises founded the journal Zeitschrift für Angewandte

Mathematik und Mechanik. This journal was the most prestigious journal in mechanics for many years.

In aerodynamics, von Mises made notable advances in boundary-layer-flow theory and airfoil design. He also made an important contribution in solid mechanics, the von Mises yield criterion for plastic flow of ductile metals.

More on von Mises is presented in the later section of this book on immigrants to the United States.

Nikoloz I. Muskhelishvili

Nikoloz I. Muskhelishvili (Tbilisi, Russia; St. Petersburg, Petrograd, Tbilisi, Transcaucasian, 1891–1976) was a Russian-applied mathematician from the Georgia region of the USSR. His most significant contribution is his extensive book on the mathematical theory of elasticity (Muskhelishvili 1933). He is known for the application of complex variables for the solution of plane elasticity problems. He was the first president of the new USSR Academy of Sciences that was established in 1941. He was also chair of the USSR National Committee on Theoretical and Applied Mechanics from 1956 to 1976.

Ivan S. Sokolnikoff

Ivan S. Sokolnikoff (Chernigov-Province, Russia; Idaho, Wisconsin, Brown, UCLA, 1901–1976) received his early education in, Kiev (Russia at the time). Following participation as a Naval Officer during the Russian Revolution, he went to China where he worked for an American electrical firm. With assistance from the Salvation Army and the American Consul in Harbin, China, he immigrated to Seattle, WA in 1921. He completed a degree in electrical engineering at the University of Idaho in 1926 and a Ph.D. in mathematics at the University of Wisconsin in 1930. He remained on the faculty at Wisconsin until 1964 when he moved to UCLA. He also completed visiting professorships at Brown, London, Brussels, Zurich and Ankara.

Sokolnikoff's mathematical specialty was elasticity theory. His treatise, *The Mathematical Theory of Elasticity* (Sokolnikoff 1956), is a classic. The textbook on mathematics for engineers and physicists, written with his first wife Elizabeth (nee Stafford) (Sokolnikoff and Sokolnikoff 1941), was for many years the leading book in the field. Sokolnikoff also wrote texts on advanced calculus and tensor analysis. He served as editor of the Quarterly Journal of Applied Mechanics and the John Wiley Series in Applied Mathematics.

Sergei Gheorgievich Lekhnitskii

Sergei Gheorgievich Lekhnitskii (Leningrad State Univ., Saratov Univ., 1909–1981) was born on June 22, 1909 in the Yaroslavl region of Russia, roughly 170 miles north of Moscow and 500 miles south of St. Petersburg. He studied physical and mathematical sciences with a specialty in mechanics at the Leningrad State University (now St. Petersburg State University) completing his degree in 1931. He then did graduate studies at Leningrad in Fracture Mechanics working with Professors S. A. Gershgorin and G. V. Kolosov and completing his graduate courses in 1934. It was during his graduate studies that he first worked on the theory of anisotropic elastic bodies. He developed the application of analytic function of complex variables to problems of stress concentration around holes in anisotropic plates, and torsion and bending of anisotropic rods.

He stayed at Leningrad University teaching and conducting research, rising to a position of senior researcher of the Institute of Mathematics and Mechanics. In 1937, he accepted a newly established chair of elasticity theory at Saratov State University (in Saratov Russia). Lekhnitskii remained at Saratov until 1959 when he returned to Leningrad as the senior research fellow.

During his time at Saratov, he submitted his doctoral thesis entitled "Some problems in the theory of elasticity of an anisotropic body" and received the degree Doctor of Physical and Mathematical Sciences in 1941 from Leningrad.

Lekhnitskii was a leading scholar in the theory of anisotropic elasticity. He published several books on anisotropic elasticity including *Stability of Anisotropic Plates*, and *Torsion of Anisotropic and Non-Homogeneous Bars*; researchers working in the field of fibrous composite materials consider his book *Theory of Elasticity of an Anisotropic Body* (Lekhnitskii 1950) to be a classic.

Lekhnitskii died in August 1981 at age 72.

John von Neumann

John von Neumann (Budapest, Hungary; Berlin, Hamburg and Princeton, 1903–1957) was an Hungarian-born applied mathematician who became an American citizen in 1937. At birth his name was János Lajos Neumann. He later added the von to his name (the title had been granted to his father) and used John rather than János. It was recognized at a very young age that he had exceptional ability for memory and mathematics. He obtained a degree in chemical engineering at the Technische Hochschule in Zurich and a doctorate in mathematics from the University of Budapest, both in 1926. He lectured at Berlin from 1926 to 1929 and at Hamburg from 1929 to 1930. He did postdoctoral work under Hilbert at Göttingen during 1926–1927. It was at Göttingen that he did the early development of his work on quantum mechanics. He joined Princeton University in 1930 and remained there in the Institute for Advanced Study until his death in 1957.

von Neumann made major contributions in many areas including functional analysis, numerical analysis, quantum mechanics, fluid mechanics, computing, and linear programming. He was a key member of the Manhattan Project that developed the first atomic bomb at Los Alamos, New Mexico, USA.

In his autobiography, von Kármán tells an interesting story about von Neumann. During von Kármán's first year at Aachen, a well-known banker from Budapest (no

small distance, it is over 750 miles) came with his 17-year-old son Johnny. The
banker wanted von Kármán to dissuade his son from studying mathematics. von
Kármán suggested a compromise, the son should study chemical engineering and
would also have courses in mathematics. Johnny did go to Zurich and study chemical engineering, but fortunately for the world, he switched to mathematics and
became the father of the digital computer.

In many ways, von Neumann was much like his fellow Hungarian von Kármán.
He was very sociable hosting many large parties in his large home at Princeton, was
a poor driver but liked to drive, and enjoyed off-color humor.

He died in Princeton at age 53 after an 18-month bout with cancer.

Warner Tjardus Koiter

Warner Tjardus Koiter (Amsterdam, The Netherlands; Delft, 1914–1997) was born
in Amsterdam on June 16, 1914. After primary and secondary education, he entered
Delft University of Technology in 1931 and earned a degree in mechanical engineering. During his final year at Delft he was assistant to Prof. C. B. Biezeno, one
of the founders of the International Applied Mechanics Congresses. From 1931 to
1938, he worked on aircraft structures at the Dutch National Aeronautical Research
Institute (NLL) in Amsterdam. Two years were spent at the Government Patent
Office and the Government Civil Aviation Office.

During the Second World War Koiter was allowed to work at NLL on subjects of
his own choosing. This work led to his Ph.D. thesis, *On the Stability of Elastic
Equilibrium*, which was awarded by Delft University in November 1945 with
Biezeno as the advisor. Koiter's thesis actually was completed and approved by
Biezeno in 1943. The plan was to have the thesis printed with an oral exam in the
fall of 1943. However, in August 1943 the Nazis required a declaration of loyalty to
the Nazi occupation authorities from all students who wished to pass an examina-
tion. Unwilling to do this, Koiter delayed printing of the thesis and, therefore, completion of his degree, until after the war. Koiter had a strong dislike for the occupying

German forces; his brother had died in a concentration camp. His thesis stability work was written in Dutch and was not widely known until it was translated into English in 1960.

Koiter was one of the longest serving member of IUTAM (1952–1986). He served on the IUTAM Bureau from 1956 to 1972, and as President from 1968 to 1972. His honors include the ASCE von Kármán Medal in 1965 and the ASME Timoshenko Medal in 1968. As Chair of the ASME Applied Mechanics Division in 1996, I had the privilege of writing to Koiter to inform him that the Warner T. Koiter Medal had been established in his honor and that he was to be the first honoree.

Koiter gave a retrospective of his career in *Forty Years in Retrospect, the Bitter and the Sweet* in the book Trends in Solid Mechanics (Besseling and Van Der Heijden 1979). He was a member of the Royal Netherlands Academy of Sciences, the Royal Society, and the National Academy of Engineering (NAE).

Chapter 3
Major Developments in Mechanics

Mechanics at Göttingen

The University of Göttingen in Göttingen, Germany had a tradition of outstanding mathematicians for much of the nineteenth century, beginning with Carl Friedrich Gauss, Professor at Göttingen from 1807 to 1855. Johann Peter Gustav Lejeune Dirichlet succeeded Gauss in 1855; he was succeeded by George Friedrich Bernhard Riemann in 1859. Riemann taught at Göttingen until shortly before his death in 1866. Felix Klein arrived at Göttingen in 1886 and re-established Göttingen as the leading mathematics center in the world at the turn of the century. There can be no question that Klein's emphasis on applied mathematics at Göttingen played a fundamental role in the establishment of mechanics as a separate discipline of engineering science.

Klein hired the mathematician David Hilbert in 1895 and the engineer (mechanician) Ludwig Prandtl in 1904. While Prandtl may have been better known and may have attracted more students to study at Göttingen, many physicists, mathematicians, and mechanicians credit Klein with instilling in them an appreciation for the application of mathematics as an engineering science called *mechanics*. In addition, Klein was one of the first academics to develop a close working relationship with industrial and governmental organizations, an approach followed later by a number of leading universities in the United States including Caltech, Stanford, and MIT.

Well-known mechanicians who spent time at Göttingen include: Theodore von Kármán, Stephen Timoshenko, Arpad Nadai, J. P. den Hartog, Jerome Hunsaker, William Prager, Richard Courant, and Sydney Goldstein. Other well-known scientists and mathematicians who spent time at Göttingen early in the twentieth century include: Robert Millikan, Paul Ehrenfest, Dirk Struik, David Hilbert, Max Born, Robert Oppenheimer, Enrico Fermi, Edward Teller and John von Neumann.

© Springer International Publishing Switzerland 2016
C.T. Herakovich, *Mechanics IUTAM USNC/TAM*,
DOI 10.1007/978-3-319-32312-1_3

Mechanics generally was not the primary research focus of all of these men, but they often knew each other quite well and spoke the same language having had Göttingen as a common experience. This list of Göttingen-influenced mechanicians and other scientists grows exponentially when the students of these scholars are considered.

Clearly, Göttingen was the epicenter for the growth and development of the engineering science of mechanics in the early twentieth century.

International Research Council

The years following World War I (after 1918) were difficult for scientists who wanted to collaborate with scientists in all other countries. In some cases, strong animosities had developed between the warring factions. Following overtures from Germany to initiate peace negotiations, President Woodrow Wilson of the United States proposed that international cooperation in science be re-establish, but with the specific condition that Germany and her allies be excluded. Wilson proposed an International Research Council organized around International Unions, one for each of the various scientific disciplines. His proposal was accepted in 1919 and committees were formed in 11 countries of the Allied Powers. The International Unions had the power to invite neutral countries to join, *but not those countries against whom the Allied Powers had fought*. Germany, Austria, Hungary, and Bulgaria were not permitted to be members under the terms of the agreement. Thus, their scientists could not attend official international meetings.

The Innsbruck Conference

The establishment of the U.S. National Committee on Theoretical and Applied Mechanics (USNC/TAM) grew, in large part, out of a desire by U.S. mechanicians to participate in the ongoing (European) International Congress of Applied Mechanics. European mechanicians were involved in joint activities in the form of international mechanics meetings dating from 1922 when what is now referred to as the *Innsbruck Conference* was convened.

In 1921, Prof. Theodore von Kármán was in his tenth year as the Director of the Aeronautical Institute at the Techniche Hochschule in Aachen, Germany. He lived in a large house in Vaals, Holland (just across the border from Aachen) with his mother and sister. Mother and sister looked after von Kármán's social life leaving him to concentrate on his intellectual pursuits. The large house was the site of many social gatherings. In particular, on weekends, the house was open to students, assistants, and visitors. It was common for people speaking a wide variety of languages to be present where they consumed much food and wine in a very gay atmosphere.

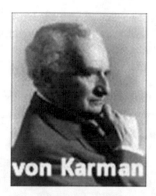

These social gatherings were very successful and, according to von Kármán's autobiography, he credits his sister Josephine (Pipö) for suggesting one day "that they do something to maintain regular contact with scientists in other countries". von Kármán attributes the desire for continual international interactions by him and his sister to their father who had taught them to have a worldview, and the fact that they were "Hungarians living on Dutch soil and working in Germany".

Because of the approval of International Research Council as proposed by President Wilson, it was not an easy time to arrange an international conference among European scientists. It was only 3 years after the First World War. The War, with primarily Germany, Austria-Hungary, Bulgaria, and the Ottoman Empire (the Central Powers) on one side and France, the United Kingdom, Italy, the Russian Empire, Japan and the United States (the Allies) on the other, had left many scientists with strong feelings. After the war, German scientists had lost communication with the British, French, and Americans; the French had excluded Germany from all international congresses.

Details of behind-the-scenes process leading up to the Innsbruck conference are provided in an article entitled "*The Early International Congresses of Applied Mechanics*" by G. Battimelli of Rome in Stephen Juhasz's "*IUTAM: A Short History*" (Juhasz 1988). Battimelli reviewed much of the correspondence between von Kármán and others that preceded the Conference. What Battimelli writes is consistent with the briefer account that appears in von Kármán's autobiography "*The Wind and Beyond*" (Kármán with Edson 1967). Being an astute politician, von Kármán realized that it was important that one of the organizers of a conference involving scientists from both sides of the conflict be a respected scientist from one of the Allied Powers. Italy had joined the Allies, so, in April 1922, von Kármán wrote to Professor Levi-Civita, a highly respected mathematician in Rome, asking for his advice and collaboration on the organization of an *informal* conference on hydro- and aeromechanics. Kármán believed that the time was ripe to break the dependence of the fluid mechanicians from the established disciplines such as mathematics and to give hydro- and aeromechanics the independent status they deserved. Since the conference was to be "informal," scientists from all countries could be invited.

Levi-Civita, who was known for his pacifist and internationalist views, was enthusiastic in his response and suggested that they also invite some English and French scientists. von Kármán wrote to Stodola in Zurich "political views should be completely bypassed by the fact that the meeting will not represent any *official* Congress, but will be held as a totally informal gathering". Unfortunately Stodola declined to participate as did H. A. Lorentz, the leading Dutch scientist, Vito Volterra, "*Mr. Italien Science*", and Richard von Mises from Berlin. Several French scientists sent message of sympathy and adhesion, but did not attend.

von Kármán's suggestion for an informal conference on mechanics was driven in part by the desire that he, and others, wanted to establish applied mechanics as a separate field, distinct from applied mathematics and physics. During and after a 1920 scientific meeting in Nauheim, Germany, leaders including Prandtl, von Kármán, von Mises and Trefftz (see Battimelli) had agreed that separate sessions for applied mechanics were needed at these scientific meetings. Separate session for applied mechanics did appear at the meeting of this scientific group a year later.

The results of the sessions on applied mechanics were prominently reported in a new (February, 1921) journal Zeitschrift für angewandte Mathematik und Mechanik (*Journal of Applied Mathematics and Mechanics*) edited by von Mises in Berlin. This journal was to become a primary reference for scholarly papers on mechanics for many years. According to Battimelli, von Kármán summed up the "connection between applied goals and theoretical research, empirical work, and mathematical investigation: "turning engineering design into engineering science"".

The letters of invitation eventually were signed by von Kármán, Levi-Civita, Prandtl and C. W. Oseen, a Swedish physicist. A group of 30 Europeans met in Innsbruck, Austria, in September 1922 for the informal conference on hydro- and aeromechanics. von Kármán and his sister paid all secretarial expenses using their own money. Pipö attended and assisted with the meeting, as she often would throughout her life. The list of attendees included many of the outstanding scientists in the field. They came from Germany, Austria, Holland, Scandinavia, and Italy. The meeting was a great success; it set the stage for the International Applied Mechanics Congresses that were to follow.

International Applied Mechanics Congresses

During the Innsbruck Conference, von Kármán suggested that an International Congress covering all areas of applied mechanics be held at a future date. J. M. Burgers from The Netherlands was an enthusiastic supporter. As a result, the First International Congress of Applied Mechanics was held in 1924 at the Technical University of Delft, The Netherlands. Delft was selected as the place for the meeting, in part, because it was a neutral country during the War. The organization of the 1924 Congress was entrusted to Profs. J. M. Burgers (chair of fluid mechanics) and C. B. Biezeno (expert in elasticity theory) at Delft. Working with von Kármán, they developed a ten-member executive committee that was as international as possible, and re-affirmed that the conference was to be totally informal. The French declined invitations to be members of the executive committee.

Two hundred and seven scientists attended the Delft congress where 76 papers were presented. The delegates all wore badges reading IMC, which stood for International Mechanics Congress, and, as von Kármán said, was also an apt expression for cooperation in several languages. He said, one delegate put it, "IMC in German stands for "Ich muss cooperieren," in English, "I must cooperate," and in French, "Isolation me coûtera" (Isolation will cost dearly)."

During the Delft Congress, the executive committee decided to establish The International Congress for Applied Mechanics as an ongoing activity. A permanent International Congress Committee (replacing what was formerly called the executive committee) was formed to decide upon general matters related to future congresses including the selection of host countries. It was emphasized that the International Congress Committee members were to "act only as individuals and by no means represented any institution; no official designation by Academies or governmental agencies would be accepted, and the committee would only grow by co-optation". The membership of this first International Congress Committee included two Americans (J. S. Ames and J. C. Hunsaker) and two future Americans (R. von Mises and Th. von Kármán).

During the 1924 Delft Congress, it was decided that the organization of each subsequent congress was to be entrusted to a National Committee of scientists from the host country. There continued to be problems deciding on the location of future meetings because of strong nationalist feelings. However, over time, and with diplomacy, these feelings were assuaged. After 1924, and prior to the Second World War, International Congresses were held in Zurich, Switzerland (second, 1926), Stockholm, Sweden (third, 1930), Cambridge, UK (fourth, 1934), and Cambridge, Massachusetts, USA (fifth, 1938).

Following the 1938 Congress, the Second World War interrupted the convening of international congresses until 1946. The Sixth International Congress convened in 1946 in Paris, France. It was at this meeting that the International Union of Theoretical and Applied Mechanics (IUTAM) was formally established. Prior to that time, the group of mechanicians was concerned only with the international congresses held once every 4 years. The Union was established with the goal of supporting interactions on a broader scale. Interactions would not be limited to 4-year intervals.

As of 2016, the Congress Committee of IUTAM continues to have total responsibility for deciding the host country and the technical program of future congresses. The IUTAM General Assembly elects the members of the Congress Committee. However, decisions of the Congress Committee are not subject to approval by the General Assembly. The Congress Committee selects the site of future congresses by secret ballot of its members.

The independence of the Congress Committee has an interesting history. The international congresses prior to 1946 had experienced unusual success. According to Battimelli there were two factors that contributed to the success. The "founding fathers refused to establish any formal connection between the Congress and any official body or institution tied to the International Research Council or to single Governments. This "refusal of politics" proved to be an extremely successful political act". The new discipline of applied mechanics was a scientific activity that bordered on physics, mathematics, and engineering, and it had matured in those years mainly in the German-speaking scientific world. "Proposals advanced by the International Research Council to establish international cooperation were tied to old-fashioned disciplinary subdivisions, and generally inspired by anti-German prejudice; they totally failed to materialize".

Chapter 4
Mechanics in the United States

There was relatively little science-based mechanics in the United States at the beginning of the twentieth century. However, the invention of human-powered flight by the Wright Brothers in 1903 and engineering advancements related to World War I (1914–1918) jump-started the growth of science-based mechanics. This was true initially in Europe. The interest in mechanics was due in large part to the desire for a better understanding of aerodynamics. In the United States, it was realized, somewhat belatedly, that there was an urgent need for science-based mechanics, and science-based engineering in general, if the country was to be militarily and economically competitive. The activities in mechanics played a leading role in the establishment of science-based engineering design, manufacturing, and engineering education in the United States. The science-based developments relied much more heavily on mathematics for analytical approaches to the formulation and solution of engineering problems. Prior to these developments, empirical methods (knowledge gained by observation) were the standard approach for understanding physical phenomena. The following paragraphs describe some of the major events that influenced the conversion to science-based mechanics in the United States.

Land Grant Colleges: 1862

The first official government action in the United States that can be related to mechanics was the Morrill Act for Land Grant Colleges. The Morrill Act was proposed in 1857. Congress passed an authorization bill in 1859; however, President James Buchanan vetoed it. In 1861, Morrill resubmitted the act with an amendment that the proposed institutions would teach military tactics as well as engineering and agriculture. This was during the American Civil War (1861–1865). Aided by the secession

© Springer International Publishing Switzerland 2016
C.T. Herakovich, *Mechanics IUTAM USNC/TAM*,
DOI 10.1007/978-3-319-32312-1_4

of a number of states, states that did not support the original plan, the reconfigured Morrill Act was passed by Congress. President Abraham Lincoln signed it into law on July 2, 1862.

The Morrill Act allowed for the creation of "land grant colleges" in the individual states. The stated purpose of the land grant colleges was:

> ... without excluding other scientific and classical studies and including military tactic, to teach such branches of learning as are related to agriculture and the mechanic arts, in such manner as the legislatures of the States may respectively prescribe, in order to promote the liberal and practical education of the industrial classes in the several pursuits and professions in life.

Note the use of the term *mechanic arts* in the statement of purpose.

National Academy of Sciences: 1863

The Morrill Act was followed soon after by an Act establishing the National Academy of Sciences (NAS) in 1863. President Abraham Lincoln signed this bill on March 3, 1863, also during the Civil War. This Act states:

> The Academy shall, whenever called upon by any department of the Government, investigate, examine, experiment, and report upon any subject of science or art, the actual expense of such investigations, examinations, experiments, and reports to be paid from appropriations which may be made for the purpose, but the Academy shall receive no compensation whatever for any services to the Government of the United States.

The bill to establish the NAS was introduced by Senator Henry of Massachusetts during the last hours of a congressional session when the Senate was immersed in the rush of last minute business. Without examining it or debating its provisions, both the Senate and House approved the bill and President Lincoln signed it. (see *The Lazzaroni: science and scientists in mid-nineteenth-century America*, 1972, page 10, Smithsonian Institution Press.)

As the reader may be aware, this has become a rather common tactic in the U.S. Congress. Attach a small item to a larger bill, or add it at the last minute, and it may be passed with little scrutiny. More on this approach for congressional approval of initiatives is considered later when the establishment of the National Advisory Committee for Aeronautics (NACA) is discussed.

The National Academy of Sciences is important to the U.S. National Committee on Theoretical and Applied Mechanics (USNC/TAM) because, after functioning as an independent committee of scholars interested in mechanics for 18 years (1948–1966), the committee agreed, at the invitation of the NAS, to become a committee under the umbrella of the NAS. The end-result of this happenstance is that once the USNC/TAM agreed to become a committee of the NAS, it became subject to oversight and jurisdiction by the NAS

Under the authority of its charter, the National Academy of Sciences established the National Academy of Engineering in 1964 and the Institute of Medicine in 1970. Neither entity has any direct relationship with USNC/TAM. Thus, in the eyes of the NAS, mechanics is a science-based activity as opposed to an engineering activity.

First Flight: 1903

The Wright Brothers historic first flight in 1903 had a revolutionary impact on the study of aerodynamics and science-based mechanics. An event that piqued von Kármán's interest in aerodynamics was his presence at the first two-kilometer airplane flight in Europe. It was in March 1908 at Issy-les-Moulineaux, just outside Paris. Englishman Henry Farman was to attempt a record breaking, sustained, two-kilometer flight in a Voisin biplane for a 50,000-franc purse. von Kármán was in Paris relaxing after completing his doctoral thesis (on buckling) under the guidance of Ludwig Prandtl at Göttingen. In a café on Boulevard St. Michel early one morning, he was asked by Margrit Veszi, daughter of a friend from Budapest who was writing an article for a Paris newspaper, to give him a ride to the historic event that was to take place at 5 AM that morning. von Kármán agreed to give Margrit the ride. The flight that he observed that morning had a lasting impact on his future research interests.

While still in Paris, von Kármán received an invitation from Prandtl to return to Göttingen and work with him to build a wind tunnel and conduct aerodynamic experiments. The initial plan was to investigate airship models in the wind tunnel. There was growing interest among the European science and engineering communities in the potential for airborne flight. The primary focus was on dirigibles. Planes had not yet developed to the point of general acceptance as airborne machines. There was relatively little, if any, research in aerodynamics in the United States at this time.

This was 1908; scientific research leaned very heavily on the empirical approach of laboratory investigations. However, through the leadership of Felix Klein and Prandtl at Göttingen, the marriage of science-based mechanics (mathematical formulations) with engineering was beginning to take hold. Prandtl's boundary layer work of 1904–1905 was one of the earliest examples of this marriage. In 1911–1912, von Kármán published his mathematical analysis of vortex shedding, later called the von Kármán vortex street. Much of Kármán's work was unknown outside of Göttingen until possibly 1921 when von Kármán published additional work on the boundary layer (see Anderson, *Physics Today*, December, 2005).

The first wind tunnel in the United States may have been the small one built by the Wright brothers to study the merits of a variety of wing shapes prior to their historic 1903 flight. J. C. Hunsaker , with the assistance of student D. W. Douglas, built the first wind tunnel of significant size in the U.S. in 1914 at MIT. Hunsaker took the lead building this wind tunnel after observing the wind tunnels at Göttingen in Germany and Teddington in England. The work at MIT was the first organized research activity on aeronautics in the United States. Hunsaker's contributions are discussed in more detail later. Douglas went on to be the founder of the Douglas Aircraft Company.

National Advisory Committee for Aeronautics: 1915

As early as the 1890s, scientists in the United States recognized that the U.S. was lagging behind the Europeans when it came to scientific approaches to advancements in science and engineering (see for example, Robert A. Millikan's 1950 autobiography page 19). With the approach of World War I, the nation was beginning to realize that it needed a center for aeronautical research as a means of closing the gap with the advancements taking place in Europe.

The National Advisory Committee for Aeronautics (NACA) began as an emergency measure prior to World War I to promote industry/academic/government coordination on war-related projects. President William Howard Taft had appointed a National Aerodynamical Laboratory Commission in 1912, but legislation to approve funding for the commission was defeated in congress in 1913.

Charles D. Walcott, who was then Secretary of the Smithsonian Institution, led a renewed effort to get the commission approved by congress. Assistant Secretary of the Navy, Franklin D. Roosevelt, wrote that he "heartily" endorsed funding the commission. Walcott then suggested adding funds for the commission to the Naval Appropriations Bill. The resolution stated that the purpose of the commission was:

> to supervise and direct the scientific study of the problems of flight with a view to their practical solution, and to determine the problems which should be experimentally attacked and to discuss their solution and their application to practical questions.

Legend has it that "the enabling legislation for the NACA slipped through almost unnoticed as a rider attached to the Naval Appropriation Bill, on 3 March, 1915". President Woodrow Wilson signed the bill into law on the same day creating the Advisory Committee for Aeronautics. The unpaid committee of 12 individuals was provided a budget of $5000 per year.

This event demonstrates, once again, that the manner in which congress functions has not changed much over many decades. Smaller items are included in much larger bills in order to get them funded with relatively little notice.

The first meeting of the National Advisory Committee for Aeronautics (NACA) was held in the Office of the Secretary of War on April 23, 1915. Brig. Gen. George P. Scriven was elected as the temporary Chairman, and Dr. Charles D. Walcott was elected Chairman of the NACA Executive Committee. Those in attendance at the inaugural meeting were: Dr. William Durand, Stanford University; Dr. S.W. Stratton, Director, Bureau of Standards; Brig. Gen. George P. Scriven, Chief Signal Officer, War Dept.; Dr. C. F. Marvin, Chief, United States Weather Bureau; Dr. Michael I. Pupin, Columbia University; Holden C. Richardson, Naval Instructor; Dr. John F. Hayford, Northwestern University; Capt. Mark L. Bristol, Director of Naval Aeronautics; Lt. Col. Samuel Reber, Signal Corps, Aviation Section; Dr. Joseph S. Ames, Johns Hopkins University; and Hon. B. R. Newton, Asst. Secretary of Treasury.

Walcott became Chairman of NACA and served until 1927. Walcott had been appointed to the chair position because of his involvement in securing congressional approval of the commission (as it was originally called). However, Walcott had no background in aeronautics, his field was paleontology. It appears that Ames (a physicist) was appointed Chair of the NACA Executive Committee in order to provide the desired scientific leadership.

The manner in which NACA was given congressional approval explains, in large part, how it was that J. C. Hunsaker of the Navy was assigned to MIT to lead in the development of aeronautics in the United States. Clearly, as a result of the Naval

Appropriation Bill of 1915, the Navy had a vested interest in aeronautics. These developments also provide background as to how it is that the *Navy*, primarily thought of as a sea-going activity, assumed a leadership role in the development of airplanes for warfare.

In the following years leading up to and through World War II, there was competition, and some consternation, between the Navy and the Army over which service should be responsible for the development of air warfare. Eventually, of course, both services have an air wing, and we now have the U.S. Air Force as a separate entity.

National Research Council: 1916

Another significant development in the United States related to mechanics was the establishment of the National Research Council (NRC) within the National Academy of Sciences (NAS) in 1916. The First World War started in 1914, but the United States did not enter the war until 1917. However, in 1916, prompted by concerns over the possibility of U.S. entry into the war, a resolution offered by George Hale of the NAS recommended that the NAS offer its resources to the President of the United States in the interest of national preparedness. The NAS approved the resolution. Following a meeting at the White House, President Woodrow Wilson accepted the NAS offer to "organize the scientific resources of educational and research institutions in the interest of national security and welfare" under the umbrella of a National Research Council, operating as a committee of the NAS. During the early years of the NRC, there were strong participations by the military services and engineering societies.

Following the U.S. entry into the war in 1917, President Wilson signed an Executive Order requesting the National Academy of Sciences to perpetuate the National Research Council. The Order signed by Wilson on May 11, 1918 listed the first and primary duty of the NRC as:

> In general, to stimulate research in the mathematical, physical, and biological sciences, and in the application of these sciences to engineering, agriculture, medicine and other useful arts, with the object of increasing knowledge, of strengthening the national defense, and of contributing in other ways to the public welfare.

Robert Millikan

Robert Millikan, a Nobel physicist, played important roles in the formation and the work of the NRC during and after World War I. He served as the Executive Director of the NRC from 1916 to 1919. Millikan had spent time in Europe, Germany in particular, in 1895–1896. During this first visit to Germany, he studied in Berlin and Göttingen. He studied at Göttingen for one semester. One of the courses he took while there was that of the mathematician Felix Klein . In his autobiography (page 32), Millikan makes the statement that "Klein's crusade actually gave a big boost to applied mathematics in Germany, notably in aeronautics and fluid mechanics". Millikan visited Germany again in 1912 for an extended visit. Thus, he was well aware of the scientific activities in Germany as well as the buildup of the armed forces prior to World War I.

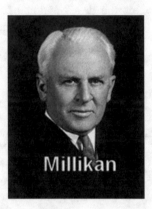

Millikan was also important to the establishment of the USNC/TAM because he invited Theodore von Kármán to visit Caltech in 1926, and eventually convinced von Kármán to become a permanent member of the Caltech faculty in 1930.

Refinements to the duties of the NRC were codified by an Executive Order of President Dwight D. Eisenhower on May 10, 1956 and further by President George H. W. Bush on January 19, 1993 (the day before he left office).

Chapter 5
Immigrants to United States

The developments in science-based mechanics in the United States are due in large part to the immigration, in particular, of two outstanding mechanicians from Europe. Stephen P. Timoshenko arrived from Kiev, Ukraine, Russia, by way of Zagreb, Yugoslavia, in 1922, and Theodore von Kármán arrived from Budapest, Hungary by way of Aachen, Germany (part time starting in 1926, permanently 1930). These two men knew each other well. They both had studied at the University of Göttingen in Germany working on stability under the guidance of Ludwig Prandtl. Timoshenko was in residence at Göttingen from April to September, 1905, for the summer of 1906, and again from April to September 1909. von Kármán arrived in Göttingen in October 1906 and was there continuously, as student and then *privat dozent* (effectively instructor), until September 1912. While at Göttingen, both men took classes from mathematician Felix Klein and were influenced by Klein's strong belief in the importance of applied mathematics, which was a rather new concept at the time. As Director of the group at Göttingen, Klein initiated chairs in applied mathematics and applied mechanics. He also established a close working research relationship between university and industry; a model that eventually would be followed by American universities such as Caltech, Stanford and MIT.

As a result of their experiences at Göttingen, Timoshenko and von Kármán developed a strong belief in a science-based approach to mechanics in particular, and to engineering in general. A major tenet of this approach was mathematical formulations for the representation of physical phenomena. Both men completed their dissertation work on stability under Prandtl who served as von Kármán's major advisor and as advisor number 2 for Timoshenko. However, it was during this time that Prandtl became more interested in aerodynamics and von Kármán's postdoctoral work with Prandtl was in aerodynamics.

Prior to their arrivals in the United States, both men were known in Europe for their outstanding contributions to mechanics. However, when Timoshenko arrived in the United States his previous work was virtually unknown in the United States. In contrast, von Kármán's work was known when, in 1926, he was recruited by Robert Millikan to develop a program in aeronautics at Caltech.

© Springer International Publishing Switzerland 2016
C.T. Herakovich, *Mechanics IUTAM USNC/TAM*,
DOI 10.1007/978-3-319-32312-1_5

Stephen P. Timoshenko (1922)

Timoshenko immigrated to the United States from Zagreb, Yugoslavia (now Croatia), where he had worked for 2 years. He lcft the Ukraine in 1920 because of the unstable political conditions that existed there during the previous decade or more. He arrived in the United States in June 1922, 3 months prior to the September 1922 international mechanics conference that von Kármán had initiated in Innsbruck. It is unlikely that Timoshenko would have been invited to the Innsbruck meeting as he had not worked in fluid mechanics and that conference was devoted entirely to hydrodynamics and aerodynamics. In the United States, he quickly developed a reputation and a following for the outstanding courses he taught in solid mechanics and vibrations, his books on these subjects, and the excellent work he did at Westinghouse.

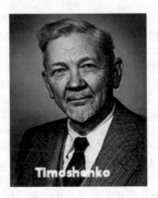

He was responsible for starting the Applied Mechanics Division of ASME and, as an indirect result, the U.S. National Committee on Theoretical and Applied Mechanics. Timoshenko's role in the development of activity in mechanics in the United States can be attributed to his leadership in promoting science-based mechanics education in the United States. He arrived in the United States to begin work at a small company in Philadelphia that specialized in vibration problems (known as the Vibration Specialty Company). He was recruited to join the company by a former student (Zelov) of his at St. Petersburg Polytechnic. The company was owned by another Russian engineer by the name of Akimov. After working in Philadelphia for less than a year, Timoshenko accepted a position in Pittsburgh with the Westinghouse Company. The Vibration Specialty Company was not doing well, and he became aware that Westinghouse was in the process of expanding their research in mechanics. The "wedding" of pure science to engineering applications had always interested him. He now had the opportunity to do this on a grand scale.

As indicated above, he immigrated to the United States in 1922 and soon was employed by Westinghouse. He was 44 years of age when he arrived In America; his previous work was virtually unknown by the Americans. He stayed at Westinghouse for 5 years, and then moved to the University of Michigan in 1927 (at age 49) where he stayed for 9 years. He moved to Stanford University in Palo Alto, California in June 1936; he was 58 years old when he joined the Stanford faculty.

At Westinghouse, Timoshenko taught courses on mechanics in the Westinghouse "design school". A school the company organized for some of the young engineers that Westinghouse was hiring—often 300 or more new hires in a year. It was Timoshenko's belief that the young American engineers were lacking seriously in the basics of engineering science, and mathematics in particular. His Westinghouse work and the courses were very successful and word spread about their high quality. During his time at Westinghouse Timoshenko also made presentations at several American universities (Illinois, Michigan and MIT) as well as scientific (mathematics) and engineering (ASME) meetings in the United States. While at Westinghouse, he proposed that ASME initiates a section on applied mechanics. George Eaton of Westinghouse, and an ASME member, helped enormously in this effort. ASME officially established the Applied Mechanics Division in 1928. More on this new division will follow in a later section.

Things had gone so well at Westinghouse that the company agreed to send him to Europe for 2 months, July 15—September 15, in 1926. He had requested this trip so that he could attend the 1926 International Congress of Applied Mechanics in Zurich (his first International Applied Mechanics Congress). He also attended a convention of the British Association for the Advancement of Science at Oxford, and made visits to several industrial testing laboratories. Returning to Westinghouse, he was given permission to work at home in the morning for several days a week to write a book on *Vibration Problems in Engineering*. He completed the book in the spring of 1927.

Some of Timoshenko's associates during the years at Westinghouse (1923–1927) included men who would later become university professors and play leading roles in the development of mechanics in the United States. Examples include J. Ormondroyd, mechanics professor at the University of Michigan, G. Karelitz, professor of applied mechanics at Columbia, J. P. den Hartog, professor of mechanics at MIT and C. R. Soderberg, professor and Dean at MIT.

As a result of his growing reputation, he was recruited to the University of Michigan for the 1927–1928 academic year. He gladly accepted this offer as it would give him more quiet time to work on his writing. Also, it gave him the freedom to travel to Europe in the summer, without the need of approval by higher authority, nor the submission of a trip report. At Michigan, he created the first bachelor's and doctoral programs in engineering mechanics in the United States. During the 1929–1930 academic year, he taught the first university course in elasticity. His time at Michigan may be known best for the Summer School on Applied Mechanics. The courses taught in the summer schools were attended by young professors from universities across the United States; they were very successful. In addition to the formal courses of the Summer School, a series of lectures by local and invited guests was organized. The guests included men such as Prandtl, von Kármán, R. V. Southwell, G. I. Taylor and J. M. Westergaard.

Many of the summer school students returned to Michigan to complete a doctorate working with Timoshenko. Twenty-nine students completed doctorates under

Timoshenko at Michigan. Many of them became well-known in mechanics. Examples include: L. H. Donnell, M. M. Frocht, J. N. Goodier, G. H. MacCullough, E. E. Weibel, D. H. Young, and M. I. Hetenyi.

In 1936, Timoshenko moved to Stanford where he established an applied mechanics program similar to the one at Michigan. Nine students received doctorates under him at Stanford. The more well-known are: E. H. Lee, N. J. Hoff and E. P. Popov. Thus, the total number of doctoral students directed by Timoshenko in the United States was 38.

Throughout his life, Timoshenko spent most of his summers in Europe, writing, relaxing, and visiting friends. Sections of his autobiography (Timoshenko 1968) read like a travelogue of Western Europe. He officially retired from Stanford in 1944, but continued lecturing there until the spring of 1955 as a "professor-at-large", a position that was created for him. He also returned to Michigan for the Mechanics Summer School during some of these years.

Timoshenko moved to Wuppertal, Germany in 1964 to be with his daughter and her family. His wife had died in 1946 in Palo Alto. Timoshenko died on May 29, 1972, in his daughter's home, at age 93. He died after a brief illness from a kidney ailment. His ashes were buried in Palo Alto. He had become a citizen of the United States in 1927; however, he makes no mention of this in his autobiography.

On page 317 of his autobiography, Timoshenko makes the statement "In the 40 odd years of my wanderings I have visited and lived in many countries, but only in Yugoslavia have I never felt foreign. The 2 years that I lived there [Zagreb] appear to me now as the happiest of my life." A most interesting statement in view of all the success he had while in the United States.

Among his many awards and recognitions, the most highly recognized is the ASME Timoshenko Medal established and awarded to him, in 1957. The ASME Applied Mechanics Division is responsible for selection of the awardee. He received the ASME Worcester Reed Warner Medal (1935), ASME Honorary Member (1952), ASEE Lamme Medal (1939), Levy Medal (1944), and Cresson Medal (1958) from the Franklin Institute. International Medals included: Grande Médaille (France), James Watts International Medal (Great Britain), Trasenter Medal (Belgium), James Ewing Medal (Great Britain), Jourowski Medal (1911) and Salov Prize (1945) Russia. He was elected to the National Academy of Sciences (1940), Ukrainian Academy of Sciences (1918), Russian Academy of Sciences (1928), Polish Academy of Technical Sciences (1935), French Academy of Sciences (1939), Royal Society, London (1944), and Italian Academy of Sciences. Timoshenko was Chair of the ASME Applied Mechanics Division in 1927 and 1930. In addition, eight institutions in seven different countries conferred honorary doctoral degrees.

Timoshenko was the author of five books in Russian and 13 books in English. Many books were also translated in other languages. His books and his style of teaching may be his most important contributions that shaped the field of applied mechanics education during the twentieth century.

Timoshenko Books

Timoshenko's influence on solids mechanics is evident from the following list of his books written in English. As many of these books had black covers, students at Stanford often referred to them as Timoshenko's "black books". If each edition and each translation is counted as a single book, the books that Timoshenko wrote in English and were then translated into a language other than English total to the amazing number of 105. The five books he wrote in Russian prior to arriving in the United States also had many editions. His autobiography provides a complete listing.

- *Applied Elasticity* (with J. M. Lessells), Van Nostrand, 1925
- *Vibration Problems in Engineering*, Van Nostrand, 1st Ed. 1926, 2nd Ed. 1937, 3rd Ed. 1955 (with D. H. Young)
- *Strength of Materials, Part I*, Van Nostrand, 1st Ed., 1930, 2nd Ed. 1940, 3rd Ed., 1955
- *Strength of Materials, Part II*, Van Nostrand, 1st Ed., 1930, 2nd Ed. 1941, 3rd Ed., 1956
- *Theory of Elasticity*, McGraw-Hill, 1st Ed. 1934, 2nd Ed. 1951 (with J. N. Goodier)
- *Elements of Strength of Materials*, van Nostrand, 1st Ed., 1935, 2nd Ed., 1940, 3rd Ed., 1949 (all with G. H. MacCullough), 4th Ed. 1962 (with D. H. Young)
- *Theory of Elastic Stability*, McGraw-Hill, 1st Ed. 1936, 2nd Ed. 1961 (with J. M. Gere)
- *Engineering Mechanics* (with D. H. Young), McGraw-Hill, 1st Ed., 1937, 2nd Ed., 1940, 3rd Ed., 1951, 4th Ed. 1956
- *Theory of Plates and Shells*, McGraw-Hill, 1st Ed., 1940, 2nd Ed., 1959 (with S. Woinowsky-Krieger)
- *Theory of Structures* (with D. H. Young), McGraw-Hill, 1st Ed., 1945, 2nd Ed., 1965
- *Advanced Dynamics* (with D. H. Young), McGraw-Hill, 1948
- *History of Strength of Materials*, McGraw-Hill, 1953
- *Engineering Education in Russia*, McGraw-Hill, 1959
- *As I Remember*, Van Nostrand, 1963 (in Russian), 1968, English translation by Robert Addis, D. Van Nostrand Company, Inc. 1968, Princeton, New Jersey

Theodore von Kármán (1930)

As noted previously, Theodore von Kármán, while at Aachen, Germany, proposed the first international conference on mechanics. The meeting was held in Innsbruck, Austria, in September 1922. It was devoted to aerodynamics and hydrodynamics and was a very successful conference. This conference led to the First International Congress of Applied Mechanics 2 years later (1924) at Delft, The Netherlands. It was at this Congress that the International Congress Committee was established. von Kármán was a member of the International Congress Committee from its inception in 1924, until his death in 1963. He was Secretary (with J. C. Hunsaker) for the 5th International Congress in Cambridge, USA. von Kármán also actively participated in meetings in the United States in 1948 and 1949 that led to the establishment of the USNC/TAM.

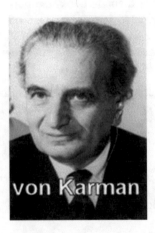

von Kármán visited the United States for the first time in 1926, at the invitation of Robert Millikan, President of California Institute of Technology, and Harry Guggenheim, President of the Daniel Guggenheim Fund for the Promotion of Aeronautics. Millikan had met von Kármán at an international physics congress in Europe in 1924. (See Dryden, it appears that Gorn and others are mistaken about where they first met. Millikan is not on the list of attendees at the 1924 International Congress of Applied Mechanics.) Millikan was impressed by the accomplishments and ability of the 43-year-old director of the Aerodynamics Institute at Aachen. In 1926, Millikan sent a telegram to von Kármán asking him to come to Caltech. He wanted him to give advice on the establishment of aeronautical courses and design of wind tunnels, and to lecture at American universities where courses in aeronautics were to be established. About the same time, von Kármán was invited to Japan for 6 months to establish the country's first major research laboratory in aeronautics. He decided to combine the two trips and took a 1-year leave of absence from Aachen.

von Kármán arrived in the United States (with his sister Josephine, Pipö) in September 1926. After visiting with Daniel Guggenheim on Long Island for a weekend, they traveled to California. He spent several weeks at Caltech assisting in the design of a wind tunnel and completing preliminary plans for the Aeronautics Laboratory. He and his sister then traveled back across the United States seeing the sights, such as the Grand Canyon, and he giving lectures at the University of Michigan, New York University and MIT (Note that Timoshenko would have been at Michigan at this time, but neither von Kármán nor Timoshenko mention seeing each other at Michigan that fall). From MIT, they went to Dayton, Ohio where they had lunch with Orville Wright and visited his private laboratory. von Kármán refers to the visit with Wright as the "peak event of this part of my visit to the USA." (see his autobiography, page 128). In December, von Kármán accompanied Pipö to New York where she boarded a ship to sail back to Europe. von Kármán then went on to San Francisco from where he set sail to Japan.

In 1928, von Kármán agreed to a position with associate status at Caltech. During the next 2 years, he split his time between Caltech and Aachen. In 1930, von Kármán settled permanently in the United States, becoming the full-time Director of the Guggenheim Aeronautical Laboratory of the California Institute of Technology (GALCIT) and Director of the Daniel Guggenheim Airship Institute at Akron, Ohio. Despite his love for Aachen, the political events in Germany and in particular the rising anti-Semitism persuaded him to accept the permanent position at Caltech. His mother and sister accompanied him to California.

The family settled in a rather large home on South Marengo Avenue in Pasadena. As in Vaal, this home became the site of many meetings, and parties with students, colleagues, and luminaries. von Kármán was a very social individual who loved to have parties and meetings in his home. He also thoroughly enjoyed a good joke, and had an eye for pretty women. A wide cross-section of people including students, Hollywood folks, Roman Catholic priests, artists, military men and academics attended the parties. Pipö had much to do with who attended the parties; she became a Roman Catholic and befriended those in the film community. Pipö earned a doctorate in art history in Vienna and was fluent in many languages.

von Kármán became a U.S. citizen in 1936. His mother and sister looked after him and his home. His sister was his manager and host until her death in 1951. Brother and sister were devoted to each other and her death plunged von Kármán into deep depression for several months during which he was unable to work. von Kármán never married.

During his time at Caltech, von Kármán made contributions in compressible flow, supersonic aerodynamics, and turbulence. He consulted for a wide range of industry and government. He played a leading role in the development of rocketry, and formed a group to build high altitude rockets and jet-assisted take-off of airplanes. The group eventually became the Jet Propulsion Laboratory. von Kármán had 38 doctoral students, the great majority of whom earned their degree at Caltech. He took emeritus status from Caltech in 1949.

In 1944, von Kármán began a series of leaves of absence from Caltech to work with several advisory groups for the U.S. war effort. He developed a very close relationship with General H. H. (Hap) Arnold of the U.S. Air Force. This resulted in von Kármán organizing and chairing a Scientific Advisory Group. The Advisory Group wrote two major reports, "Where We Stand" that described the status of various technical fields in Europe, and "Toward New Horizons," that recommended future directions for the U.S. Air force. He mostly lived in Washington, DC during this time, until 1951. After Washington, and for the remainder of his life, his home was in Paris, France, where he was Chairman of the NATO Advisory Group for Aeronautical Research and Development (AGARD). He had proposed this group be formed after the war in order to facilitate the international exchange of scientific developments.

von Kármán received many awards during his life; too many to list all of them here. He was a member of the National Academy of Sciences (1938). In 1963, at age 81, von Kármán was the recipient of the first National Medal of Science, bestowed in a White House ceremony by President John F. Kennedy. von Kármán was recognized, "For his leadership in the science and engineering basic to aeronautics; for his effective teaching and related contributions in many fields of mechanics, for his distinguished counsel to the Armed Services, and for his promoting international cooperation in science and engineering." When President Kennedy presented the medal to him, von Kármán "pledged his brain as long as it lasted" to the country of which he had become a citizen in 1936.

Some of his other U.S. awards include: ASME Medal (1941), Wright Brothers Memorial Trophy (1954), Daniel Guggenheim Medal (1955), U.S. Medal of Freedom, and ASME Thurston Lecture (1950), Timoshenko Medal (1958). According to Dryden, von Kármán was Honorary Fellow, Honorary Member or Fellow of approximately 40 national professional societies in 11 countries, received more than 60 honorary degrees or special awards from institutions in 13 international countries, and as many as 10 honorary degrees from the U.S. universities.

An excellent, detailed account of von Kármán's life was provided by his friend, Hugh Dryden, In Dryden's 1965 *Biographical Memoir of Theodore von Kármán* was published by the National Academy of Sciences. von Kármán's autobiography *The Wind and Beyond* (1967) was completed by Lee Edson after von Kármán's death. Michael H. Gorn published a biography *The Universal Man*, *Theodore von Kármán's Life in Aeronautics* in 1992.

von Kármán died on May 7, 1963, in Aachen, Germany, where he had arrived a month earlier to recuperate and rest up for a heavy travel schedule planned for the coming summer. Barbel Talbot, an old acquaintance (and widow of friend George Talbot) from early Aachen days was his "steady companion, romantic interest, and travel mate" (Gorn, page 132) during his days in Paris, and was with him in Aachen when he died.

A Theodore von Kármán Stamp was issued on August 31, 1992.

von Kármán Books

* *General Aerodynamic Theory*, 2 vols. (with J. M. Burgers), Verlag von Julius Springer, Berlin, 1924.
* *Mathematical Methods in Engineering*, (with M. A. Biot), Translated into French, Spanish, Portuguese, Italian, Turkish, Japanese, Polish, and Russian. McGraw-Hill, New York, 1940.
* *Aerodynamics: Selected Topics in Light of Their Historical Development*, Translated into Spanish, Italian, German, French, and Japanese. Cornell University Press, Ithaca, New York, 1954.
* *Collected Works of Dr. Theodore von Kármán*, 4 vols. Butterworth Scientific Publications, London, 1956.
* *From Low-Speed Aerodynamics to Astronautics*, Pergamon Press, London, 1961.

Notes on Timoshenko and von Kármán

There can be no question that Timoshenko and von Kármán both made significant contributions to the development of mechanics in the United States, and worldwide. They were of similar age with Timoshenko being 2 years older. Timoshenko made his contributions in solid mechanics and vibrations. von Kármán made his contributions in solid mechanics, fluid mechanics, and physics. Timoshenko devoted much of his effort to writing books, whereas von Kármán devoted much of his effort to collaborating with industry and governmental (mostly military) organizations, and the organization of national and international groups for scientific collaboration, including IUTAM and AGARD. Of course, both were also excellent professors.

On page 288 of his autobiography, Timoshenko recalls visiting "von Kármán, whom I knew well from my days at Göttingen" at Aachen in the summer of 1928. He goes on to relate that von Kármán took him for a drive through the city to look at ancient buildings and that von Kármán was an inexperienced driver who drove very fast. From Aachen, Timoshenko went on to see Prandtl at Göttingen during the 1928 trip. He was there to find a Prandtl student in hydrodynamics who would be interested in joining Westinghouse in America. The result of the visit was that O. G. Tietjens accepted a position at Westinghouse.

A second entry in Timoshenko's autobiography (page 311) relates being with von Kármán at the 1933 World's Fair in Chicago. He refers to von Kármán as his *old acquaintance,* and how they "talked of the persecution of the Jews by the Nazis. Having known each other for a long time, we could discuss the question with complete frankness." A third entry in Timoshenko's biography (page 326) recalls that after completion of lectures at the University of California at Berkeley in February, 1935, he (and his wife) went to Caltech where von Kármán, who had arranged their visit, met them at the railroad station and drove them to the faculty club. Clearly, they were good friends who knew each other quite well.

It is interesting that whereas Timoshenko mentions von Kármán as a friend several times in his autobiography, von Kármán only mentions Timoshenko once in his autobiography, and that is only to include him as a fellow member of a committee. However, as will be demonstrated later when Frank Marble's oral history is reviewed, von Kármán also considered Timoshenko to be a good friend.

Timoshenko was married and had three children, two girls and one boy. von Kármán never married, but had a close relationship with his sister, Pipö, and mother who lived with him and took care of his household and managed his daily affairs. His sister usually traveled with him. Later in life, after both his mother and sister had died, von Kármán developed a close relationship with Barbel Talbot, an old acquaintance from Aachen.

After immigrating to the United States, Timoshenko returned to Europe (typically Switzerland) most of his summers where he would rest, work on his writings, attend some meetings, and visit friends, mostly at universities. On the other hand, von Kármán usually remained in the United States working with his students, associates, and military organizations until 1951. In that year, von Kármán took up residence in Paris where he served as Chairman of AGARD, the group that he had established. He maintained his residence in Paris until his death in 1963; he died while on a visit to Aachen. He was in Aachen resting for a planned busy summer of scientific meetings. Barbel Talbot was at his side in the hospital when he died on May 6, 1963. This was 5 days before his 82nd birthday. von Kármán's body was flown back to the United States aboard a military aircraft. A memorial service, attended by scores of students, colleagues, scientists, and military men, was held in von Kármán's home at 1501 South Marengo Avenue. The service was performed by a Catholic priest and included a tribute from President Kennedy. He was laid to rest in Hollywood Memorial Cemetery where many movie celebrities are buried. A rabbi conducted the service. His body was placed in a simple crypt below the bodies of his sister and his mother.

Timoshenko also spent the last years of his life in Europe. He moved to Wuppertal, Germany in 1964 to be with his daughter and her family. Timoshenko died in his daughter's home on May 29, 1972, at age 93. His ashes were buried in Palo Alto alongside those of his wife.

Recollections of Timoshenko and von Kármán

Dan Frederick, former professor and chair of the Department of Engineering Science and Mechanics at Virginia Tech, responded as follows when I asked him about the course he took from Timoshenko at the University of Michigan:

Regarding Timoshenko, my memories are the following. I took only one graduate course from him in the summer of 1949 at the U. of Michigan on the theory of plates. He was an excellent lecturer in the European style who started and ended his lectures on the bell. His work on the blackboard was orderly, exact and error-free. He gave clear explanations of difficult concepts. In bearing, he was distinguished and noble. It was a great honor to have studied under one of the world's greatest scholars and professors in the field of mechanics.

When I asked Norm Abramson, former Executive Vice President of Southwest Research Institute and former chair of the USNC/TAM, about Timoshenko, he responded as follows:

I have a copy of the book {As I Remember}. I would not like to dispose of it, but I would be willing to lend it to you for a specific period of time. It means something special to me, as I was fortunate to have had somewhat of a personal relationship with Timoshenko during my four years at Stanford. I was a student in three of his classes and because I was tasked to shepherd the Flugge's and Klotter on their arrival and beginnings at Stanford, he seemed to take some special interest in me which he expressed pointedly in a personal discussion when I departed Stanford.

Your write-up on Timoshenko is very good and quite complete. I do not believe that he ever applied for U.S. citizenship, retaining his outlook and customs (way of life) as a European. He was always dressed in suit and tie, appearing very distinguished and with a rigid bearing. At the same time, he was always courteous, friendly, smiling, and interested in the students. I know that he cherished his time at both Michigan and Stanford and so it is difficult to understand the impression that he leaves with the reader in the latter parts of his book. He certainly decried the events leading to the Russian revolution as well as the rise of the Nazi government, although he was a great admirer of Germany and its education system. He was so proud of his German education that he felt that his children should remain there for their education in spite of world events. Of course, because of his strong attachment to things European, he traveled there as often as he possibly could and at the end of his active career he returned to Germany for the remainder of his life. His many trips to Europe, at least once each year, normally involved lectures at the many universities with which he was well acquainted through faculty. I am sure that he was often given rather generous travel allowances and honoraria by those institutions, and while he had favorite hotels for relaxation I am also quite sure that his many friends and acquaintances were generous in hosting him. I don't know of others who might have known Timoshenko personally, although I am sure that there are some—all of the faculty and students that I knew there are gone, with only one exception. He was my fellow student, Eugene A. Ripperger (a student of Goodier), retired Prof at UT-Austin, but he is almost 100 years old and somewhat infirm. Carl, I apologize for the rambling discourse but you have stirred up a lot of memories!!

Abramson did not know that Timoshenko had become a U.S. citizen, but his comment demonstrates that Timoshenko never advertised the fact that he had become a citizen.

Jim Simmonds, a colleague at the University of Virginia, related (to me) an interesting anecdote about Warner Koiter meeting von Kármán at von Kármán's hotel in Paris in 1946. In the spring of 1972, Simmonds was a NATO postdoctoral fellow working with Koiter at Delft. As Koiter told the story to Simmonds, when the first post-war IUTAM conference was being organized for 1946, Koiter was asked to deliver (from Delft) some papers to von Kármán who was staying in Paris. Koiter met von Kármán at the door of his hotel room about three in the afternoon. von Kármán was in a smoking jacket and remarked to Koiter (possibly in German): "This is the life. To be paid by the Americans, and to live in Paris." This is an excellent example of the jocular von Kármán.

Frank Marble on von Kármán

Frank Marble of Caltech provided interesting, and at times frank, insight on von Kármán in Shirley Cohen's interview of Marble for the Caltech Oral History Project. The interview was conducted in seven sessions January to March, 1994 and April, 1995. Marble was a close friend and associate of von Kármán from 1947 until von Kármán's death in 1963.

Marble had arrived at Caltech in September 1946 to study for a Ph.D. on an NRC pre-doctoral fellowship. He went there with the anticipation of working with von Kármán, but found that von Kármán was away a good portion of the time (including when Marble arrived) involved in war-related activity in Washington. As a result, Marble did his Ph.D. under the direction of Hans Leipmann.

However, von Kármán did return to the campus, if only intermittently, within a year after Marble's arrival in Pasadena. Marble and von Kármán developed a very close working relationship as well as a strong friendship that lasted throughout the remainder of von Kármán's life.

According to Marble, when von Kármán returned to Caltech:

> von Kármán was "holding court" with people who were allowed to come to see him. Leipmann had arranged for me [Marble] to see him. I took thorough advantage of it, and had a great time. I can remember showing von Kármán some of my work. And he said to me, "Young man, are you a mathematician?" I said, "No, Professor von Kármán, I'm an engineer. But I like mathematics." He said, "Oh, good."
>
> Well, I didn't know it at the time, but mathematicians usually occupied a level lower than engineers in von Kármán's mind. He thought they were not aware of what the real physical issues or physical problems were. So he was a little bit afraid that I was too attracted to mathematics. Well, if I had been, that certainly changed my attitude—that reaction of Kármán's. But that was the start of a very close association and friendship with Kármán. Every time he came back, he knew he could turn things over to me if he wanted them done; they would get done. He showed me a piece of work he'd done that had to do with turbulence. And, he said, "I would like you to check this over, and maybe see if you can solve this or that. Which I did; I took the thing home and worked on it and brought it back in a couple of days, and he was enchanted. And that led to von Kármán's great warm heart embracing Ora Lee [Marble's wife] and me both."

In 1949, Marble received a cable from the Air Force telling him that travel orders were being cut for him to join Professor von Kármán in Paris at an International Congress of Astrophysics and Gas Dynamics. He received the cable only 2 weeks before the trip was to commence. He was to fly on military air transport. All expenses were to be covered by the Air Force. von Kármán had set it all up without ever mentioning it to Marble.

After the Congress, Marble was to fly out of Frankfurt and von Kármán decided that Frank should travel with him (and his sister Pipö who always traveled with him), by chauffer-driven auto, as von Kármán was being driven to Heidelberg. It was a several-day trip, with a stopover in Bern, Switzerland, where von Kármán had business related to getting his brother out of Hungary. Marble sat in the back seat with Pipö who, Marble said, had a very loud and penetrating voice, spoke rapidly, and in a very domineering way. In the end, he and Pipö got to know each other well and "hit it off all right".

Marble also makes the statement that "although Kármán did not have the highest opinion of him [Timoshenko], he considered Timoshenko a good friend. So the relationship between Kármán and Timoshenko was that Kármán liked him, saw him as a good friend, but did not think of him as a first-rank engineering scientist." Another example of von Kármán's attitude about Timoshenko is revealed in another story Marble tells. Duncan Rannie brought his thesis to von Kármán, Kármán looked through it and said, "Oh, Duncan, this is quite good. This would do great credit to a Timoshenko student." As a consequence of that remark, "Duncan went back to his office and dropped the thesis into the wastebasket."

Such may have been the relationship between two outstanding scientists who, at least from von Kármán's perspective, were in competition. There seems to be little doubt that von Kármán developed a much closer personal relationship with his students and co-workers than did Timoshenko.

Marble's association with von Kármán continued through the years as von Kármán would return to Pasadena from time to time and then as a member of AGARD after it was established. In 1952, von Kármán called Marble to a meeting at his hotel in New York with Jean Fabri from France and Brian Mullins from England. von Kármán laid out his plans. He was close to the NATO organization and wanted to strengthen it by building an organization of young scientists from all the NATO countries for the purpose of developing common research aims. Kármán's idea was that people in the organization would conduct unclassified research and would exchange results as well as people. Kármán was quoted as saying, "If the youth of the world cannot cooperate in scientific things, where is our civilization going to be 30, 40 years hence?" This was the basis upon which AGARD was formed.

Marble related another story that demonstrates the worldwide respect that von Kármán commanded.

The first general meeting of AGARD was in Rome, in December 1952. In fact, it was a meeting of NATO and AGARD; it was the start of the AGARD organization. Now Kármán was in charge of the entire thing, with a positive control over it that defies description, really. Without seeming to try, he just orchestrated everything that happened. For example, with regards to the combustion panel, he got hold of me before we sat down in the general assembly to nominate folks, and he said, "Now, Frank, when I ask, or when you are asked, for nominations for the chairman of the combustion panel, you will nominate Dr. Jean Surugue, from France." I said, "Dr. von Kármán, I don't have the remotest idea who Jean Surugue is." He said, "Nevertheless, you will nominate Jean Surugue." So, came time, the questions were asked, I got up and nominated Jean Surugue as chairman of the combustion panel. Of course, that cemented the friendship of Jean Surugue and Frank Marble, even though we'd never known each other before that moment; he thought it was a gesture of great confidence and friendship. Surugue was a very fine man and played a significant role in my attachment to AGARD.

Kármán explained his reasons to me somewhat later. He said, "You know, Frank, most of the money and the organization of this entire thing comes from the United States. And the impression the Europeans get is that we are manipulating them. It must be that Europeans are the chairmen of most of the panels. They must have that right; they must have the influence to push things the way they want them." So that was his reasoning. And it worked very, very well. As usual, Kármán was a marvelous judge of what was right and what was wrong in situations of this sort.

Another anecdote that Marble tells about von Kármán is something that took place just prior to the opening of that AGARD General Assembly in Rome.

… we were all ready to march into the hall, up the stairs into the hall for the general assembly, and the Italians had put a very handsome soldier on each side of each step. So we went through this file of bright-colored military folk—just handsome and tall and strong. I walked with Kármán, as several of us did. And it was clear that he was a little bit ill at ease about this, as we started up the first step. He didn't take stairs all that rapidly in those days, anyway. So he walked up a couple of steps, and he sized the thing up as being a little too stiff, a little too formal. So he walked over to the second soldier and shook hands with him. And that just cracked everybody up—that this big stuffy guard would have this little American come up to him and shake hands with him, while they were trying to put on the biggest show of their lives with all this window-dressing. It just cracked the whole place up. And it really injected a note of levity into the entire general assembly meeting. He knew exactly how to break the ice, and that was the way he did that.

Marble also talked about von Kármán's impact on the discipline of engineering at Caltech.

Aeronautics was a place of its own. And I think it's important to say that Kármán, when he came, lifted the scientific quality of aeronautics above anything else in the division of engineering, including electrical engineering. He just had that way about him of doing it, and he carried the aura of a great European. He also was extraordinarily good at extracting money from a wide variety of people—even, in some cases, Robert Millikan. So he kept things going very well.

There developed, at that time, a bit of a split between aeronautics and the other portions of engineering, at least civil engineering and mechanical engineering, in the sense that these were old-time engineers—that is, experimentalists, largely empirical by nature. And here was Kármán, with his analytical approach and his scientific understanding of why things happen, not just what happened. And Kármán had introduced a whole new attitude of engineering. He called it engineering, but it was very different from what was practiced by any of the others. And it's one thing that made Caltech unique at that time. The aeronautics school was a unique place. Probably there's only two institutions that had anything like that, that I know of— Brown University was one of them, and that's because it went way out of its way to import— or to give refuge to, I should say—to some of the outstanding scholars that had left Europe before the war, and even during the war. Brown developed a school of applied mathematics that was second to none in the country. And, of course, Richard Courant went to NYU, and established a similar school there. It's interesting to think that all of these people were related, in the sense that their lives were influenced by Felix Klein. Felix Klein was primarily a mathematician at Göttingen, and he influenced this entire generation of people, including Kármán, Courant, and the large group of people that went to Brown. Klein, in spite of his mathematical competence and mathematical understanding, realized the limitations of pure mathematics. He acknowledged the fact that mathematical disciplines originated in physical problems, really, and that you abstracted something from the physical problem, you abstracted a mathematical description of it, and then you studied the mathematics of that. And if the mathematics got too far from the physical issues, it became a little irrelevant, he felt. He was very much interested in real problems—in physical problems, engineering problems, and such—and he instilled this in everybody that came close to him. And even though these people were not formally his students, the influence he had was phenomenal. In a way, he is the godfather of this entire new generation of engineering.

One final anecdote from Marble's oral history once again shows the devilish nature of von Kármán.

In 1949, when I was with von Kármán in Paris for this first meeting on astrophysics and gas dynamics, Lee DuBridge [Caltech President] was in Europe, too. And he had written Kármán, or had called Kármán, and said he'd like to have lunch with him on a certain day. And Kármán said to me, "Now, Frank, you know what DuBridge wants to do. He wants to retire me. I do not particularly like that idea." So Kármán, with the manner he usually had — something clever — agreed he would have lunch with DuBridge and that they would meet at a certain place at a certain time, which they did. But Kármán proceeded to invite ten other people, including me. So the lunch was at a restaurant with a large table, almost a conference-room table, with DuBridge at one end, Kármán way down at the other end, and five other people on each side who were interested in very different things, talking to Kármán and talking to DuBridge, and talking to each other. So, it turned out that Lee had to get a plane out of there before he ever really had a chance to talk to Kármán about retiring. But, of course, eventually it came to pass. Kármán retired, but not enthusiastically.

Marble was a member of AGARD from its formation in 1952 until 1965. During that time, he would travel 2–4 times a year to the NATO countries as the meetings were almost always held in Europe. Marble interacted with von Kármán at all of these meetings until von Kármán died in 1963.

Citation: Marble, Frank E. Interview by Shirley K. Cohen. Pasadena, California, January–March, 1994, April 21, 1995. Oral History Project, California Institute of Technology Archives. Retrieved 7/11/2014 from the World Wide Web: http://resolver.caltech.edu/CaltechOH:OH_Marble_F.

Other Immigrants to the United States

In addition to Timoshenko and von Kármán, a number of other immigrants played major roles in the development of mechanics in the United States. Individuals who were involved early on and deserve mention are Dirk J. Struik (from The Netherlands), Richard von Mises (from Austria–Hungary), and Arpad L. Nadai (from Austria–Hungary), Y. H. Ku (from China), Jan Burgers (from The Netherlands), and Sydney Goldstein (from Great Britain). Brief background statements on these six individuals follow.

There is a rather long list of immigrants from Europe who played leading roles in the USNC/TAM. The list includes: J. P. den Hartog, Bruno Boley, William Prager, Nick Hoff, Miklos Hetenyi, E. H. Lee, François Frenkiel, Ronald Rivlin, Paul Naghdi, Andy Acrivos, Hassan Aref, Wolfgang Knauss, Ted Belytschko, Nadine Aubry, J. D. Achenbach and Robert McMeeking. Details of their contributions are presented in the sections on the USNC/TAM.

Dirk J. Struik

Dirk Struik, a Dutch mathematician, was born in Rotterdam in 1894. In 1912 he entered University of Leiden, where he showed great interest in mathematics and physics, and was influenced by the eminent professors Paul Ehrenfest and Hendrik Lorentz. He was a close friend of Jan Burgers.

In 1924, funded by a Rockefeller fellowship, Struik traveled to Rome to collaborate with the Italian mathematician Tullio Levi-Civita. In 1925, thanks to an extension of his fellowship, Struik went to Göttingen to work with Richard Courant compiling Felix Klein's lectures on the history of nineteenth-century mathematics.

He attended the 1924 Congress in Delft and wrote a brief report describing the Congress. In 1926, he accepted a research position at MIT and, in 1928, became a member of the MIT mathematics faculty where he stayed until his retirement in 1960. He continued to do scholarly work until his death at 106 in the year 2000. Struik was educated very broadly in mathematics and physics. His scholarly work in differential geometry was widely acclaimed as was his *A Concise History of Mathematics (1987)*. Struik's book was of enormous value when writing the early history of mechanics, and it clearly shows the close association between applied mathematics and mechanics.

It seems that the presence of Struik at MIT contributed to the fact that in 1938 the Fifth International Congress was held jointly at Harvard and MIT in Cambridge, Massachusetts. Struik is listed as an attendee at the Congress, but he did not present a paper.

It is noteworthy that Struik had close associations with many of the same people and places as had Timoshenko and von Kármán i.e., Burgers, Ehrenfest, Courant, Klein, Prandtl, Levi-Civita and Göttingen.

Richard von Mises

Richard von Mises was an outstanding mechanician from Austria–Hungary who eventually immigrated to the United States. Mises knew von Kármán well and was opposed to the 1922 international conference with attendees coming from countries on both sides of the First World War.

In 1930, von Mises was a member of the International Congress Committee on Applied Mechanics as a representative of Germany. With the rise of the National Socialist (Nazi) party to power in 1933, von Mises felt his position threatened. He then moved to Turkey, where he held the newly created chair of Pure and Applied Mathematics at the University of Istanbul. He attended the 1938 International Applied Mechanics Congress in Cambridge, USA, where he was listed as a representative from Turkey.

In 1939, he immigrated to the United States and accepted a position at Harvard University where he was appointed Gordon McKay, Professor of Aerodynamics and Applied Mathematics. In 1948, von Mises was a Member-at-Large of the inaugural USNC/TAM.

He died suddenly in 1953 after completing only three chapters of his book *Mathematical Theory of Compressible Fluid Flow*. The book was completed by his wife Hilda Geiringer and G. S. S. Ludford, and published in 1958 (Academic Press).

Arpad L. Nadai

Arpad Nadai is another example of a European immigrant who came to the United States and had an impact on mechanics. Upon the recommendation of Timoshenko, Nadai, at the time, head of the applied mechanics laboratory at Göttingen, was offered a position at Westinghouse in 1927. He accepted the offer and stayed at Westinghouse for the remainder of his career. In the summer of 1930, again on Timoshenko's recommendation, Nadai led the summer school of mechanics at the University of Michigan while Timoshenko was in Europe.

Relatively little has been published on Nadai's personal life. He was born in Budapest, Austria–Hungary, in 1883. He did undergraduate work at the University of Budapest, and then received his doctorate in 1911 from the Technical University of Berlin. In 1918, he joined the Institute of Applied Mechanics at the University of Göttingen, where he was a colleague of Ludwig Prandtl. Nadai was promoted as professor at Göttingen in 1923, and became the Head of the applied mechanics laboratory as Prandtl's interests had turned to aerodynamics.

Nadai was a pioneer of the theory of plasticity, and wrote the first book on plasticity theory which was translated into English in 1931. He received the ASME Worcester Reed Warner Medal (1947) and the Timoshenko Medal (1958), the Elliott Cresson Medal of the Franklin Institute (1960), and the SOR Bingham Medal (1952). In 1975, the American Society of Mechanical Engineers established the Nadai Medal for materials scientists in his honor.

Again, we see a common thread between Nadai, von Kármán and Timoshenko. Nadai was from Budapest and did his first studies there, as was the case for von Kármán, all three spent time at Göttingen, and Nadai and Timoshenko were together at Westinghouse.

Yu Hsiu Ku

Y. H. Ku was a Personal Member of the IUTAM General Assembly from 1948 until his death in 2002. He was elected a Member of the General Assembly, from the Bureau of Education, Shanghai, P. R. China, for the 1948–1952 term. In 1950, Ku left China to become professor of electrical engineering at MIT. He thus became a U.S. member of the IUTAM General Assembly and a member of the USNC/TAM in 1950. He moved from MIT to the University of Pennsylvania in 1952 and remained there until his death in 2002. He was a member of the IUTAM Congress Committee from 1978 to 2001. IUTAM Annual Reports indicate that Ku was an active member of the General Assembly attending most meetings until his health deteriorated. His 52 years of service to the USNC/TAM and 54 years of service to IUTAM both appear to be records for longevity.

Ku was born on December 24, 1902 in Wushi, Jiangsu Province, China. He entered the Tsing Hua School in Beijing, China at the age of 13. After graduating from Tsing Hua School (later named National Tsinghua University), he received a special scholarship to study electrical engineering at MIT. At MIT from 1923 to 1928, he was awarded the Bachelor, Master and Doctor of Science degrees in electrical engineering. He completed all three degrees in three and one-half years, a record at the time; he also had the unique distinction of being the first Chinese student awarded a doctoral of science degree (ScD) from MIT. Ku returned to China in 1928 and remained there until 1950.

While in China, he held a number of academic and administrative positions: Professor, Department Chair, Dean, Director of the Aeronautics Research Institute, Director of the first Electronics Research Institute, Principal Deputy Minister of Education, and President of China National Central University. Ku was the Education Commissioner of the Shanghai Municipal Government (1945–1947). During this period, he was an adjunct professor and taught electrical engineering courses at the National Jiaotong University in Shanghai. There he taught Jiang Zemin, a future President of the People's Republic of China; they would have a unique lifelong relationship which had a significant impact on U.S.–China and China–Taiwan cross-strait relationships. From 1947 to 1949, he was the president of National Chengchi University in Nanjing. Ku had a close relationship with President Chiang Kai-shek who was the only one to occupy the university presidency before Ku. He also had been friends with Zhou Enlai during the Sino-Japanese war, when they were members of China's cabinet. Clearly, Ku was a unique individual.

Ku left China in 1950 (when the communists took over) to accept a position as professor of electrical engineering at MIT. In 1952, he moved to the University of Pennsylvania. He retired from the University of Pennsylvania in 1972 as Professor Emeritus of Electrical Engineering and Systems Engineering. He was internationally recognized for his contributions in the areas of electrical energy conversion, nonlinear systems, nonlinear controls, and boundary-layer heat transfer.

Ku received numerous honors including 28 honorary degrees: M.A. and L.L.D., University of Pennsylvania (1972); Gold Medal, Ministry of Education (1960); IEEE Lamme Medal (1972); Gold Medal, Chinese Institute of Electric Engineering (1972); Gold Medal Pro Mundi Beneficio, Brazilian Academy of Humanities (1975), IEEE Third Millennium Medal (2000); and honorary member, United Poets Laureate International.

In addition to his engineering and administrative work, Ku earned distinction in the literary field by writing many essays, novels, plays, and poems that were well-received. His complete literary works were published in 12 volumes in 1961. Ku died on September 9, 2002 at age 99.

Johannes (Jan) M. Burgers

Jan Burgers met von Kármán in 1921 and was one of the attendees at the informal Innsbruck Conference in 1922. During discussions with von Kármán at the conference he became an enthusiastic partner in the goal of establishing a broader conference covering all areas of applied mechanics. Burgers was only 4 years younger than von Kármán. He had been a student of Paul Ehrenfest at Leiden in Holland. Along with C. B. Biezeno he hosted the First International Congress of Applied Mechanics at Delft in 1924. He continued to be intimately involved with the international congresses throughout the 1920s and 1930s.

Burgers showed up at the 1946 Congress with a plan to set up a new, more permanent organization, a Union that would provide the possibility of activities between congresses and would make possible cooperation with other existing Unions. After serious discussions among the Congress Committee members, it was decided that Burgers would draft a set of Statutes that would be discussed, in particular with the British who had some reservations, before being submitted to the full committee. The draft was modified to satisfy the British and on September 26, 1946 the draft Statutes were unanimously approved.

Burgers was elected Acting Secretary of IUTAM for 1947 and Secretary for 1948–1952. He was elected a Personal Member (Member-at-Large) in 1948 and was re-elected at 4-year intervals until his death in 1981. Burgers moved to the University of Maryland in 1955, thereby becoming a member of the USNC/TAM from that time forward.

Jan Burgers was born in Arnhem, The Netherlands (Holland), on January 13, 1985. His early education was in Arnhem city schools. He then entered the University of Leiden completing his Ph.D. in physics in 1918. He was a student of Paul Ehrenfest and also took courses from and worked with H. Lorentz. He credits Ehrenfest with instilling in him a spirit of scientific inquiry. He credits fellow student and close friend Dirk Struik with help understanding the significance of mathematics.

His first academic appointment was as chair of aero and hydrodynamics at Delft University of Technology. He is responsible for Burger's equation in fluid mechanics, Burgers vector in dislocation theory, and Burgers material in viscoelasticity. He was honored with the ASME Medal in 1965 and the APS Otto Laporte Award in 1974.

Burgers died on June 7, 1981 in Washington, DC at age 86.

Sydney Goldstein

Sydney Goldstein, an applied mathematician from Great Britain, was an elected member of the inaugural IUTAM General Assembly in 1948 for the 4-year term 1948–1952. In 1950 he moved to the Technion in Israel and was listed in the 1951 IUTAM Annual Report as an elected member from Israel for 1952–1956. However, in 1955 he moved to Harvard and was listed as an elected member from the United States for 1956–1960 in the 1957 IUTAM Annual Report. He became a member of the USNC/TAM in 1955 when he moved to Harvard; he remained a member of both the USNC/TAM and IUTAM until his death in 1989.

Goldstein was born on December 3, 1903, in Kingston upon Hull, England. He attended school at Bede Collegiate School in Sunderland and then entered the University of Leeds to study mathematics with Selig Brodetsky. Realizing that Goldstein had exceptional mathematical talents, Brodetsky advised him to move to Cambridge to study mathematics. After 1 year at Leeds, Goldstein moved to St John's College, Cambridge and graduated from the Mathematical Tripos in 1925. He then was awarded an Isaac Newton Studentship to continue research in applied mathematics under Harold Jeffreys, at Cambridge. He earned the Smith's Prize in 1927 and his doctorate in 1928. His Ph.D. thesis was entitled The Theory and Application of Mathieu Functions. Following his doctorate, he was awarded a Rockefeller Research Fellowship and spent a year working with Ludwig Prandtl at Göttingen on problems in fluid mechanics.

In 1929, Goldstein split his time between being a fellow of St John's College, Cambridge and a Lectureship in Mathematics at the University of Manchester. At Manchester he followed the well-known Osborne Reynolds and Horace Lamb. In 1931, he moved to Cambridge and took over the editorship of the series Modern Developments in Fluid Dynamics upon Lamb's death in 1934. During World War II Goldstein worked on boundary-layer theory at the National Physical Laboratory. At the end of the war, he was appointed to the Beyer Chair of Applied Mathematics in Manchester.

Goldstein was a strong supporter of the State of Israel and, in 1950, he accepted a position as professor of Applied Mathematics and chairman of the Department of Aeronautical Engineering at the Technion in Haifa. The President of the university died shortly after his arrival and Goldstein was under strong pressure to assume that position. He declined the presidency but eventually agreed to serve as Vice President for academic affairs. He is credited with designing the academic framework, writing the academic constitution, and establishing the organizational structure of the Technion. He found all of this administrative work a huge burden and, in 1955, he accepted a position as Gordon McKay, Professor of Applied Mathematics at Harvard.

He was recognized for his work on steady-flow laminar boundary-layer equations, turbulent resistance to rotation of a disk in a fluid, and aerodynamics.

Goldstein's honors include: Royal Society of London (1937), Royal Netherlands Academy of Sciences and Letters (1950), Finnish Scientific Society (1975), Honorary Fellowships of St. John's College, Cambridge (1965), Weizmann Institute of Science at Rehovot, Israel (1971), Royal Aeronautical Society (1971), Institute of Mathematics and its Applications (1972), honorary doctorates by Purdue University (1967), Case Institute of Technology (1967), Technion (1969) and Leeds University (1973). He received the Weizmann Prize for Science (1953) and ASME recognized his contributions with the Timoshenko Medal in 1965.

Goldstein retired from his Harvard chair in 1970. He died on January 22, 1989, in Cambridge, MA.

Chapter 6
ASME Applied Mechanics Division

The development of mechanics as an engineering science in the United States can be closely associated with the development of the Applied Mechanics Division (AMD) of the American Society of Mechanical Engineers (ASME). As mentioned previously, Timoshenko first proposed the idea of a mechanics group or section within ASME soon after joining Westinghouse in 1923. This was after he had presented his first paper at an ASME meeting, the 1922 ASME Winter Annual Meeting. He proposed the idea of a mechanics group or section to a number of influential members of the society; however, at that time, his idea did not gain traction.

In 1926 Timoshenko attended his first (the Second) International Congress of Applied Mechanics in Zurich, Switzerland. He was quite surprised to find that he was one of only four from the United States who attended the Congress. This disturbed him and resulted in a renewed effort to start an Applied Mechanics Section within ASME. In a letter dated November 24, 1926, Timoshenko and John M. Lessells (also from Westinghouse) sent a letter proposal for an ASME Applied Mechanics Section to G. M. Eaton, Chief Engineer at Westinghouse and a prominent member of ASME. The letter outlined the reasons for the Applied Mechanics Section. See Naghdi, *A brief history of the Applied Mechanics Division of ASME*, (JAM, vol. 46, 1979) for a history of the first 50 years of the division. The article provides details of the Timoshenko proposal and the approval process. The first announcement that a new division was under consideration referred to the Division on Mechanics, Physics, and Applied Mathematics. This title was deemed too long and changed to *The Applied Mechanics Division* (AMD). The ASME Council voted to approve the Division during a meeting in October 1927.

Clarence E. Davies was the Managing Editor of the ASME magazine *Mechanical Engineering* at this time. He later became Secretary of ASME (the top administrative position), and, for many years, was a strong supporter of AMD. The formation of the Applied Mechanics Division and the related developments of papers published in ASME Transactions and presented at ASME meetings, resulted in the community gaining a solid footing and recognition in the United States. More than

© Springer International Publishing Switzerland 2016
C.T. Herakovich, *Mechanics IUTAM USNC/TAM*,
DOI 10.1007/978-3-319-32312-1_6

200 members registered in the Division prior to its formal approval by the ASME Council. As a result of this increased mechanics activity in ASME (and possibly other societies), there was a substantial increase in the U.S. participation in the International Congress of Applied Mechanics. Whereas only four from the U.S. attended the 1926 Congress in Zurich, 25 from the U.S. attended the Third Congress (1930) in Stockholm, Sweden.

The credit for the increased U.S. mechanics activity is attributed to Timoshenko and ASME for several reasons. Timoshenko not only proposed the idea of a mechanics section, he had begun teaching mechanics courses at Westinghouse in 1923. During his first year at Westinghouse, he was asked by a group of young engineers to teach them something about elasticity. As a result, the first ever course on elasticity theory in the United States was offered to a group of about 25 Westinghouse engineers, in the evening after a full day of work. At the request of George Eaton, he also gave a course on strength of materials in the Westinghouse School of Mechanics. Even though the students in the class were engineering graduates of an American university, Timoshenko felt that their preparation was so feeble that he gave them a course usually given to sophomores in Russia.

Timoshenko, his associates and his students played leadership roles as Chairs of the ASME Applied Mechanics Division: Timoshenko (1927 and 1930), Lessells (1933), Soderberg (1937 and 1938), den Hartog (1940 and 1941), Goodier (1945), Donnell (1951), Young (1953), Hoff (1955) and Hetenyi (1957).

ASME provided space and secretarial support for meetings of the USNC/TAM during the early years of the committee. C. E. Davies served in an ex-officio position as Secretary of USNC/TAM for its first 10 years, 1948–1958. The society also published the proceedings of the U.S. Congress of Applied Mechanics during the first 25 years of the congresses.

The ASME, through its Applied Mechanics Division, appears to have been the strongest early supporter of mechanics in the United States. The Society of Rheology and the Acoustical Society of America were established in 1929, but it is not known to what extent they supported mechanics as a discipline at the time. Other organizations that did support mechanics as a discipline were formed later. Examples include: American Academy of Mechanics 1942; American Physical Society, Division of Fluid Dynamics 1948; ASCE Mechanics Division 1950; Society of Industrial and Applied Mathematics 1952; Society of Engineering Science 1963; American Institute of Aeronautics and Astronautics 1963 (a merger of the American Rocket society, 1930, and the Institute of the Aerospace Sciences, 1932); U.S. Association for Computational Mechanics 1988.

Chapter 7
International Congresses: American Involvement

Six international meetings on mechanics were held between 1922 and 1946 when the International Union of Theoretical and Applied Mechanics was formally established. Two years later in 1948, the U.S. Committee on Theoretical and Applied Mechanics was established, primarily as a vehicle for participating formally in these international conferences. As has been described previously, the International Congresses were organized and administrated by a group called the International Congress Committee. The history of participation by Americans (and future Americans) who were members of the International Congress Committee during the early years of the congresses are provided in the following list where the year is the year of election to a 4-year term:

- 1924 Delft: Ames, Hunsaker (von Kármán, von Mises, Burgers)
- 1926 Zurich: Ames, Hunsaker, Timoshenko (von Kármán, von Mises, Burgers)
- 1930 Stockholm: Ames, Hunsaker, Timoshenko, von Kármán (Burgers, von Mises)
- 1934 Cambridge, UK: Ames, Hunsaker, von Kármán, Timoshenko (Burgers, von Mises)
- 1938 Cambridge, USA: Hunsaker, von Kármán, Timoshenko (Burgers, von Mises)

The decision to hold the 5th International Congress in Cambridge, MA, was made at the 1934 Congress in Cambridge, UK, by the members of the International Congress Committee. At the time, this committee consisted of 30 members from 17 countries. A full listing of the committee members is provided in the proceedings of the 1938 Congress. It appears that several individuals from the United States were involved in the decision to host the 1938 International Congress in Cambridge, MA. They are: Th. von Kármán, S. P. Timoshenko, J. C. Hunsaker, and J. P. den Hartog.

Background on some of these individuals and their involvement is provided in the following. It should be mentioned that the (many) travels to and from Europe during the period prior to the Second World War were by ocean liner and consumed a week, or more, of travel—each way. Thus, it is likely that most trips across the

© Springer International Publishing Switzerland 2016
C.T. Herakovich, *Mechanics IUTAM USNC/TAM*,
DOI 10.1007/978-3-319-32312-1_7

Atlantic involved a month or more of total time away from home and place of work. Time aboard ship could, of course, be used to conduct new work as well as prepare lectures. And once across the Atlantic, time was available for visits with persons having similar scientific interests.

Joseph S. Ames

Joseph S. Ames of Johns Hopkins University was invited to be a member of the inaugural International Congress Committee in 1924. He accepted the invitation to be a member of the committee, but was not able to attend the 1924 (first) Congress in Delft. J. C. Hunsaker served as his proxy on the committee in Delft. Ames continued as a member of the International Congress Committee for a number of years. He was Chairman of the Organizing Committee (along with Hunsaker, Timoshenko and von Kármán) for the 1938 Congress at Cambridge, USA.

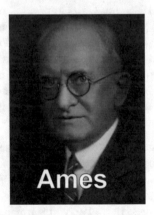

Ames was born on July 3, 1864 in Manchester, Vermont. He studied physics at Johns Hopkins University completing his undergraduate degree in 1886. His doctorate, from Hopkins in 1890, was under the advisement of the famous Henry A. Rowland, well-known for his work on diffraction gratings used in astronomical spectroscopy. After his doctorate, Ames remained at Hopkins rising through the ranks. He became a professor of physics (1889), Director of the Physical Laboratory (1901), Provost (1926) and eventually President (1929–1935). He was a founding member (1915) of the National Advisory Committee for Aeronautics (NACA, the predecessor of NASA), served as Chairman of the NACA Executive Committee from 1919 to 1936, and as Chairman of the full committee from 1927 to 1939.

The reason that Ames was invited to participate in the 1924 International Congress can be gleaned from the knowledge of his European travels prior to 1924. Following his undergraduate degree, he spent the summer traveling in Europe, and then working during the 1886–1887 academic year at the University of Berlin in Helmholtz' laboratory. His advisor Rowland had also worked for Helmholtz. Ames was in Europe again in 1900 when he presented a paper at the Congrès International de Physique in Paris. Years later in the spring of 1917, he was invited by the National Research Council (of the NAS) to head a Scientific Mission to France and England. One focus of that mission was to learn about activities related to aeronautics and the use of airplanes in warfare. As a result of these travels, and his research activities at Hopkins, Ames was known personally by a number of European scientists. It was as a result of these personal associations that he was (the only) American invited to be a member of the International Congress Committee for the 1924 Congress.

NASA Ames Research Center in California is named in his honor. He was a member of the National Academy of Sciences (1909), a Fellow of the American Academy of Arts and Sciences (1911), and was the 1935 recipient of the Langley Gold Medal from the Smithsonian Institution. Ames died on Nov. 15, 1943. An excellent, detailed review of his life is that by Henry Crew in a presentation to the National Academy of Sciences in 1944.

Jerome C. Hunsaker

Jerome Hunsaker was the only American (at the time) who actually attended the First International Congress for Applied Mechanics at Delft in 1924. Hunsaker was there as a stand-in for J. S. Ames of Johns Hopkins University who was not able to attend. In the Proceedings of the Delft Congress, Hunsaker is listed as being from London. That was because he was on assignment as a naval attaché in London from late 1923 through most of 1925. During this time, he visited many research universities and industrial facilities throughout Europe. His main interest during these visits was aeronautics. Hunsaker became a member of the IUTAM Congress Committee as a result of his participation in Delft, and his reputation in aeronautics. He remained a member of the Congress Committee for some time thereafter. Hunsaker was accompanied to Delft by his wife Alice who was listed as a member (i.e., attendee) of the Congress, as were all "accompanying ladies". Hunsaker did not present a paper in Delft.

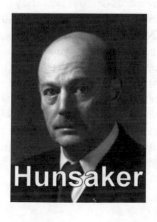

Jerome Hunsaker was born in Creston, Iowa on August 26, 1886. He graduated at the head of his class with a B.S. from the U.S. Naval Academy in 1908. In 1909, Hunsaker was assigned by the Navy to the Boston Navy Yard with a concurrent appointment to the graduate program at MIT. While there he became interested in aeronautics, in part, through his, and his wife's friendship with MIT President and Mrs. Richard C. Maclaurin. He received an M.S. from MIT in Naval Architecture in 1912.

In 1913, Hunsaker, with the permission of Gustave Eiffel (of Eiffel tower fame), published a translation of Eiffel's *The Resistance of the Air and Aviation*. Eiffel was so impressed with Hunsaker's translation that he invited him to Paris to study with Eiffel. That same year, Hunsaker, at the request of MIT President Maclaurin, was assigned by the Navy to MIT to develop an aeronautics curriculum and related research programs. There followed several months of temporary duty in Europe to survey aeronautical laboratories. He, along with Professor Albert F. Zahm of Catholic University, spent time in Pairs with Eiffel, and then visited Ludwig Prandtl at Göttingen and several other research laboratories and manufacturing facilities in Europe.

At the National Physical Laboratory at Teddington in Middlesex, England, Hunsaker became familiar with their new wind tunnel; he undoubtedly also had observed the wind tunnel at Göttingen during his visit there. Returning to MIT in 1914, he built, with the assistance of MIT student Donald W. Douglas, the first wind tunnel, of significant size, in the United States. Hunsaker's work in the wind tunnel resulted in his doctorate in Aeronautical Engineering in 1916.

Douglas was the first person to receive a B.S. in Aeronautical Engineering (1914) and remained at MIT for another year as an assistant to Hunsaker. Douglas would eventually establish the Douglas Aircraft Company. Hunsaker designed the first aircraft to fly over the Atlantic Ocean. The flying boat NC-4, flew from Newfoundland to Portugal and England in the first trans-Atlantic flight in May 1919.

It is clear that well before the 1924 International Congress of Applied Mechanics, Hunsaker had a number of acquaintances in the European mechanics community. After the 1924 Congress, he continued to be involved with the International Congress Committee, as well as with the formation of the U.S. National Committee on Theoretical and Applied Mechanics. He was Secretary (with Theodore von Kármán) of the 1938 International Congress in Cambridge, MA. He participated in meetings at ASME Headquarters in New York in 1948 where the formation of USNC/TAM was developed. These meetings also led to a proposal from the committee to become a member of the International Union of Theoretical and Applied Mechanics.

At MIT, Hunsaker was the Head of Mechanical Engineering (1933–1947), and Aeronautical Engineering, from 1939 to 1951. He was Chairman of the National Advisory Committee for Aeronautics (NACA) from 1941 to 1956. Thus, Ames and Hunsaker provided the leadership for NACA for all but 2 years (1940–1941) over the 37 years, 1919–1956. Hunsaker was followed as Chairman of NACA by Jimmy Doolittle, famous as commander of the 1942 Doolittle Raid over Japan in the early days of U.S. involvement in World War II.

Hunsaker received many honors. Among the more significant are: National Academy of Sciences (1935); National Academy of Engineering (1967); Fellow, American Academy of Arts and Sciences; Honorary Fellow, Imperial College of Science and Technology, Institute of the Aeronautical Sciences, and Royal Aeronautical Society; Honorary Member, ASME (1942). Honorary Degrees were conferred by: Williams College; Northeastern University; Adelphi College. Awards included: Navy Cross; Daniel Guggenheim Medal (1933); Franklin Medal; Presidential Medal for Merit; Legion of Honor; Wright Brothers Memorial Trophy (1951); Godfrey L. Cabot Trophy; Langley Medal.

Hunsaker retired from the MIT faculty in 1952 at age 65, but maintained a presence in his office in the Guggenheim Laboratory at MIT well into his 80s. He would walk there from his home on Beacon Hill. An excellent and detailed biography of Hunsaker is *A Biographical Memoir* by Jack L. Kerrebrock.

Jerome Hunsaker died on September 10, 1984 at age 98.

Hugh L. Dryden

The native born American who had the most impact on the formal establishment of IUTAM and USNC/TAM was Hugh Dryden. Joseph Ames and Jerome Hunsaker were involved with International Congresses on Applied Mechanics before Dryden, but they did not participate in the development of IUTAM or USNC/TAM to the extent that Dryden did.

Hugh Dryden was born in Pocomoke City, MD, on July 2, 1898. He excelled in mathematics and graduated with honors from Johns Hopkins with a B.A. in 1916, and an M.A. in Physics in 1918. Following the completion of his master's degree, he began work at the National Bureau of Standards (NBS) in their Wind Tunnel Division. He also took courses in fluid dynamics completing his Ph.D. at John Hopkins in 1919 at the age of 20. His major was physics and mathematics. He was appointed director of the newly created Aerodynamics Division of NBS in 1920 where he worked on airflow, turbulence, and boundary layer phenomenon. In this position, he undoubtedly became familiar with von Kármán's work.

Dryden had a strong interest in applied mechanics and presented papers at several International Congresses on Applied Mechanics, prior to the formation of IUTAM, including, the third (Stockholm, 1930), the fourth (Cambridge, UK, 1934), the fifth (Cambridge, USA, 1938), and the sixth (Paris, 1946).

As mentioned previously, it was at the Paris Congress that IUTAM was formally proposed. Dryden was the first IUTAM Treasurer (1948–1952), then President (1952–1956), and Vice-President (1956–1960). He took an active role in the organization of the 6th International Congress in Paris (1946) and the 8th International Congress in Istanbul (1952).

More on the contributions of Hugh Dryden is presented in Chap. 10.

Jacob Pieter den Hartog

J. P. den Hartog was born in Ambarawa on the Island of Java in the Dutch East Indies on July 23, 1901. His parents were Dutch from Amsterdam, and he returned there for his schooling. He graduated from the Technical University of Delft in 1924 with a degree in electrical engineering. There is no evidence that den Hartog attended the 1924 Congress in Delft. This is not surprising in that he had just completed undergraduate work in electrical engineering and did not become interested in mechanics until later when he worked under Timoshenko at Westinghouse in the United States.

Not finding a job in Holland after his undergraduate degree, he immigrated to the United States in 1924 where he soon was hired at Westinghouse as an electrical engineer. At Westinghouse, he took classes in mechanics from Stephen Timoshenko. Timoshenko converted him into a mechanical engineer and he became an expert in vibrations. In the evenings, den Hartog took courses in mathematics at the University of Pittsburgh earning a Ph.D. in 1929. It is somewhat surprising that while den Hartog received his Ph.D. from the University of Pittsburgh, his advisor was Ludwig Prandtl of Göttingen, Germany.

Following his Ph.D. studies, den Hartog took a sabbatical year away from Westinghouse as a postdoctoral student in Prandtl's laboratory in Göttingen. From 1932 to 1942 he was a professor at Harvard University. During this time, he was involved with the Applied Mechanics Division of ASME; the division started by his mentor Timoshenko in 1927. He also became involved with the International Congress of Applied Mechanics. He presented a paper at the 4th Congress (1934) in Cambridge, UK and was a co-editor (with Heinrich Peters of MIT) of the proceedings of the 5th Congress (1938) in Cambridge, USA. Den Hartog was a guest lecturer for several summers at the summer school for mechanics teachers hosted by Timoshenko at the University of Michigan. The closeness and admiration he felt towards Timoshenko and Prandtl is demonstrated by the name of this second son, Stephen Ludwig.

Den Hartog volunteered for a commission in the U.S. Naval Reserve in 1939 and was on active duty for 4 years 1941–1945. The last year of military service was spent in Paris as a member of a technical mission debriefing enemy technicians and securing valuable technical equipment. Den Hartog was ideal for the mission as he spoke Dutch, Flemish, German, and French, in addition to English. He was deactivated from military service in September 1945, at which time he assumed a new post as professor of mechanical engineering at MIT. His lectures at MIT were considered to be some of the best as were his books on *Mechanical Vibrations (1956)*, *Mechanics (1948)*, *Strength of Materials (1949)*, and *Advanced Strength of Materials (1952)*.

Den Hartog was Chair of the Applied Mechanics Division of ASME in 1941 and 1942. The ASME J. P. den Hartog Award recognizes *lifetime contributions to the teaching and practice of vibration engineering*. Additional awards include: ASME Worcester Reed Warner Medal (1951), Honorary Member (1964), Thurston Lecture (1970), Russ Richards Award (1947), ASME Medal (1979), and the Timoshenko Medal (1972). He was a member of the National Academy of Sciences (1953).

Chapter 8
International Union of Theoretical and Applied Mechanics

Prior to 1946, international congresses were the only activity conducted by the group of mechanicians. Professor Jan Burgers (Delft) presented a plan at the 1946 Congress in Paris to expand the activities of the group under the umbrella of a Union. A British scientist by the name of Stratton, who was general secretary of the International Council of Scientific Unions, proposed (to Burgers) that the Congress of Applied Mechanics transform itself into a Union so that it would have a more permanent character and could adhere to the International Council. This was described in a letter from Burgers to G. I. Taylor dated January 18, 1946. Taylor responded, "If he (Taylor) sees Kármán, he would ask him what he thinks". H. Villat, head of the French organizational committee for the Paris Congress, wrote to Burgers in June 1946 that several of his colleagues generally approved of the idea.

At a meeting of the Congress Committee in Paris, "it was decided to create a more permanent organization, so as to provide the possibility of carrying out activities in the interval between the congresses". In attendance at this committee meeting were: H. Villat, M. Roy, J. Pérès, A. Caquot, K. Popoff, R. V. Southwell, G. I. Taylor, C. B. Biezeno, J. M. Burgers, J. Ackeret, R. von Mises, J. P. den Hartog, and Th. von Kármán. The name of the organization was to be International Union of Theoretical and Applied Mechanics. Burgers drafted a set of statutes that the Congress Committee approved unanimously on Thursday, September 26, 1946. It appears that at least six participants from the USA were at the 1946 Congress: Th. von Kármán, J. C. Hunsaker, Hugh L. Dryden, J. P. den Hartog, R. von Mises, and Stephen P. Timoshenko.

By June 1947, ICSU (International Council of Scientific Unions) had accepted IUTAM as an adhering body. In September 1947 a provisional IUTAM Bureau (Executive Committee) was chosen with R. V. Southwell (UK, acting president), H. Villat (France, acting vice-president), Hugh L. Dryden (USA, acting Treasurer), and J. M. Burgers (The Netherlands, acting secretary). At the 1948 Congress in

© Springer International Publishing Switzerland 2016
C.T. Herakovich, *Mechanics IUTAM USNC/TAM*,
DOI 10.1007/978-3-319-32312-1_8

London, refinements to the statutes were approved and officers were elected: J. Pérès as President, Southwell as Vice-President, Dryden as Treasurer, and Burgers as Secretary.

IUTAM General Assembly

The IUTAM General Assembly includes representatives of adhering organizations (mostly countries), members-at-large (also called personal members) and an eight-member Bureau that serves as the Executive Committee. In 2012, there were 51 IUTAM adhering countries. The number of representatives of adhering countries is determined by the level of subscriptions (dues) paid by the country, based upon a sliding scale from 1 to 5 representatives. The United States is the only country that pays sufficient dues to have five representatives. The number of members-at-large is limited as a percentage of the number of members in General Assembly. In 2012, there were 118 voting members of the General Assembly. In additional to the voting members of the General Assembly, a variety of organizations are represented by non-voting observers.

IUTAM Bureau

The eight-member IUTAM Bureau is the Executive Committee of IUTAM; it is responsible for conducting business between meetings of the IUTAM General Assembly. The General Assembly meets only once every 2 years. The table in Appendix A lists the members of the IUTAM Bureau during the period 1948–2012. (Country abbreviations are those used by the International Olympic Committee.) A review of the table of Bureau Members reveals that many of the giants of mechanics have participated in the leadership of IUTAM. The Bureau was a male-only group until 2012 when the General Assembly elected Nadine Aubry of the United States to Bureau. The table is color coded for the six countries with the largest number of Bureau members. The Bureau has always had a limit that any one country could have at most one member of the Bureau. There has also been a strong effort to ensure that fluid mechanics and solid mechanics are represented equally on the Bureau. Twenty countries have had representation on Bureau as of 2012. The United Kingdom and the United States are the only two countries to have had a member on the Bureau during every year of its existence. Members of USNC/TAM have played important roles in the leadership of IUTAM starting with the involvement of Hugh Dryden as the first Treasurer of IUTAM.

Timoshenko was never a member of the IUTAM Bureau and von Kármán was not a member of the Bureau, but was elected Honorary President of IUTAM in 1951.

von Kármán-Honorary President of IUTAM

It is interesting that von Kármán was not elected as an officer of IUTAM in 1948. However, the respect and esteem for him was clearly demonstrated in 1951 when he was elected Honorary President. A section of the 1951 Annual Report of IUTAM reads as follows:

By unanimous vote of the General Assembly (recorded by correspondence) it was decided to nominate Dr. Theodore von Kamran *Honorary President* of the International Union of Theoretical and Applied Mechanics, on the occasion of his seventieth birthday, May 11, 1951, in view of his outstanding services to the science of Mechanics and of his great interest and help given in matters concerning the Union. The President of the Union, Professor Pérés, had a testimonial booklet printed with the following text:

Cher Collègue,

L'Assemblée Générale de notre Union s'associe cordialement aux voeux, aux félicitations, aux témoignages, d'admiration que vous sont addressés á l'occasion de votre Jubilé.

Nous sommes heureux de rendre hommage ici:

— au Savant dont les travaux sur l'Elasticité et la Plasticité, sur la Résistance des fluids, sur les phénomènes de Turbulence, sur l'Aérodynamique des grandes vitesses, marquent des progrès décisifs;

— au Professeur qui a créé et dirigé de puissants Centres de recherches et de qui des générations d'élèves, venus de tous les coins du monde, ont appris et apprennent ce que vaut l'Analyse mathématique, quand elle s'appuie sur un sens très aigu du réel et quand elle s'affine par le soin constant d'atteindre le but par les voies les plus droites;

— à l'Ingénieur enfin: vous avez eu un rôle de premier plan dans nombre de recherches techniques et de réalizations industrielles; votre Œuvre dormine le développement actuel de la Technique aéronautique.

L'Union Internationale de Mécanique Théorique et Appliquée vous doit beaucoup, non seulement pour la contribution de l'ordre scientifique et technique que vous apportez à nos reunions, mais aussi parce que votre activité s'est efficacement exercée, sur le plan humain, dans l'intérêt fondamental des relations scientifiques internationals. Vous aviez déjà, en 1922, é l'un des initiateurs des journées d'Innsbruck, qui préludèrent aux Congrès Internationaux de Mecanique, dont vous êtes resté un des principaux animateurs. Vous étiez présent lorsque, en 1946, notre Union s'est constituée, en liaison avec le Comité des Congrès. A ce moment, comme aussi par la suite, nous avons largement bénéficié de vos conseils que rendaient précieux votre expérience, votre largeur de vue et cette sûreté intuitive de jugement qui caractérise déjà votre oeuvre scientifique.

Notre Assemblée Générale vient, par un vote unanime, de vous élire Président d'Honneur de l'Union international de Mécanique théorique et appliquée. C'est un grand plaisir pour nous que de vous faire part de cette désignation, que nous vous prions d'accepter comme témoignage de gratitude et d'affectueuse admiration.

Dans les souhaits que nous formons pour vous, nous ne vous séparons point de votre chère soeur, dont le charme a si souvent éclairé nos réunions.

A vous, cher President, en bien cordial dévouement,

Le Bureau de l'IUTAM			
J. Pérès (Paris)	R. V. Southwell (Oxford)	J. M. Burgers (Delft)	H. L. Dryden (Washington)
F. H. van den Dungen (Bruxelles)	G. Colonnetti (Rome)	H. Favre (Zurich)	H. Solberg (Oslo)

An English translation of the testimonial provided by Monique and Jim
Simmonds reads as follows:

Dear Colleague,
The Union's General Assembly cordially joins with the good wishes, the congratulations,
the recognition, and the admiration that have been addressed to you on the occasion of
your jubilee.
   We are happy to add our praise to:
   The Scientist whose work on Elasticity and Plasticity, on viscous flows, on turbulence,
and on high speed aero dynamics has marked decisive progress in these areas.
   The Professor who has created and powerfully directed research centers and from
whom generations of students, coming from all corners of the world, have learned and
continue to learn the value applied mathematical analysis when it rests on a sharp sense of
reality and when it is refined by constant attention to the strictest procedures in attaining
its goal.
   And lastly to,
   The consummate Engineer: You have played a leading role in the development of
numerous research techniques and industrial successes; your work pervades current
technical aeronautics.
   The International Union of Theoretical and Applied Mechanics owes you much, not
only for the high scientific and technical standard you have brought to our meetings, but
also because your contributions have been efficiently and effectively exercised in a
humane way in the interests of international science. Already, in 1922, you were one of
the initiators of the days in Innsbruck, a prelude to the International Congress of
Mechanics of which you remain one of the principal advocates. You were present when,
in 1946, our Union was formed in liaison with the Committee of the Congress. At that
moment, as well as afterwards, we have benefited greatly from your council that has
distilled your experience, your overarching view, and that unfailing intuitive judgment that
has likewise characterized your scientific work.
   By a unanimous vote, our General Assembly has just elected you as the *Honorary
President* of the International Union of Theoretical and Applied Mechanics. It is a great
pleasure for us to award you this title that we fervently urge you to accept as an expression of our gratitude and affection.
   In extending to you our best wishes we do not wish to overlook your dear sister whose
charm has so often been on display at our meetings.
   To you, worthy president, in cordial devotion, we are:

Le Bureau de l'IUTAM			
J. Pérès (Paris)	R. V. Southwell (Oxford)	J. M. Burgers (Delft)	H. L. Dryden (Washington)
F. H. van den Dungen (Bruxelles)	G. Colonnetti (Rome)	H. Favre (Zurich)	H. Solberg (Oslo)

Following the testimonial, the 1951 IUTAM Annual Report went on to state:

Failure of health of Dr. von Kármán's sister, who shared so greatly in all his work,
prevented Dr. von Kármán's attendance at the formal presentation of the testimonial. It
was read and received for him at the First United States National Congress of Theoretical
and Applied Mechanics, Chicago, Illinois, June 11–16, 1951. The sudden death of Dr.
Josephine de Kármán on July 2, 1951, threw a tragic shadow over the event.
   Dr. von Kármán forwarded a telegram of thanks and appreciation to Professor Pérès.

Chapter 9
USNC/TAM

The United States National Committee on Theoretical and Applied Mechanics (USNC/TAM) grew out of interactions between the U.S. and European scientists at International Applied Mechanics Congresses, and the related establishment of the International Union of Theoretical and Applied Mechanics (IUTAM) in 1946.

It appears that von Kármán, Timoshenko and Dryden were the prime movers behind the establishment of USNC/TAM. von Kármán had initiated what was to become the International Congresses on Applied Mechanics (the forerunner of IUTAM), he arrived in the United States in 1930. Timoshenko arrived in the U.S. in 1922, developed a strong following and reputation through the courses on mechanics that he taught at Westinghouse, Michigan and Stanford; he initiated what was to become the ASME Applied Mechanics Division. Timoshenko and von Kármán were good friends both having studied under Prandtl at Göttingen during several overlapping months. von Kármán developed a particularly close relationship with Dryden; both had presented papers at the 3rd (Stockholm, 1930), 4th (Cambridge, UK, 1934), and, 5th (Cambridge, USA, 1938) International Congresses. Both men worked in fluid mechanics and aerodynamics, were joint editors of Applied Mechanics Reviews for a time, and were involved together in several governmental activities during and after the Second World War. They knew each other well and became good friends.

As described in Paul Naghdi's *A Brief History of the Applied Mechanics Division of ASME* (JAM, vol. 46, 1979), there was a close relationship between the ASME Applied Mechanics Division and those involved in the formation of USNC/TAM. A meeting to discuss the U.S. participation in IUTAM was held at ASME Headquarters in the Engineering Societies Building in New York City on January 28, 1948.

© Springer International Publishing Switzerland 2016
C.T. Herakovich, *Mechanics IUTAM USNC/TAM*,
DOI 10.1007/978-3-319-32312-1_9

In attendance were: S. Wilmot (ASCE), L. A. Burckmyer, Jr. (AIEE), H. L. Dryden (incorrectly listed as J. L. Dryden in Naghdi's report) (IAS/AIAA), S. L. Tyler and A. B. Newman (AIChE), J. R. Kline (AMS), H. A. Barton (APS), and C. E. Davies, H. W. Emmons, and S. E. Reimel (ASME).

At this meeting, an ad hoc committee was appointed to develop a plan for the U.S. participation in the Seventh International Congress for Applied Mechanics (to be held in London, England, September 5–11, 1948) with Hugh Dryden as Chair of the committee. The committee held a second meeting on April 8, 1948. Those in attendance included: H. L. Dryden, Th. von Kármán, R. R. Dexter (Institute of Aeronautical Sciences), E. Reissner (AMS), R. J. Seeger (APS), C. E. Davies (ASME Secretary and Chief Executive) and S. E. Reimel (ASME-Committee on International Relations).

At the April meeting, Dryden reported on information received from J. M. Burgers at IUTAM regarding the possibility of U.S. participation in IUTAM. It appears that the response was of a positive nature as the committee adopted the name "U.S. Committee on Theoretical and Applied Mechanics" and proceeded with plans for U.S. participation in IUTAM. At a subsequent meeting on December 2, 1948 (after the London Congress), the committee adopted the original charter of the USNC/TAM. The charter included a name change to "U.S. National Committee on Theoretical and Applied Mechanics (USNC/TAM)".

It is noteworthy that as of 1948, all activities associated with the establishment of USNC/TAM were initiated by individuals, with no formal input from societies or governmental organizations other than ASME providing meeting space and secretarial assistance. It was agreed at the December 1948 meeting that the $400 per year in IUTAM dues would be provided by contributions from the societies that were represented by members on the committee. The societies continued to provide the funds to pay the IUTAM dues until 1966 when USNC/TAM became a committee of the NAS and the Department of State paid the dues.

At a meeting on March 28, 1949, it was reported that participation was pledged by seven societies: ASME (Applied Mechanics Division), SESA, ASCE (Engineering Mechanics Division), APS (Fluid Dynamics Division), AIChE, AMS, and Institute of Aeronautical Sciences (later to become AIAA). The membership of USNC/TAM at that time consisted of 13 members: H. L. Dryden, Th. von Kármán, S. Timoshenko, J. C. Hunsaker, R. von Mises (member-at-large), H. W. Emmons (ASME), R. D. Mindlin (SESA), M. G. Salvadori (ASCE), T. B. Drew (AIChE), R. J. Seeger (APS), N. J. Hoff (IAS), E. Reissner (AMS), and C. E. Davies (ex-officio from ASME).

The USNC/TAM continued as a committee of individuals with no operational or organizational ties to a professional society or governmental organization from 1948 until 1966 when the committee agreed to accept the invitation from the National Academy of Sciences to become a committee of the NAS. During these early years, the committee was self-governing with no outside limitations or guidance from societies or government. In that sense it was, like IUTAM, a self-governing organization. Several societies did provide limited funding for IUTAM dues. The committee continued to function, very successfully, in this manner for 18 years.

USNC/TAM Timeline

The following table presents a timeline listing of major events leading up to the establishment of the United States National Committee on Theoretical and Applied Mechanics (USNC/TAM).

Year	Major USA-related mechanics developments
1922	• Timoshenko begins work (and teaching mechanics courses) at Westinghouse
1924	• 1st International Congress of Applied Mechanics held in April 22–26 at Delft, Holland
	• von Kármán attends 1st International (Delft) Congress
	• Dirk Struik (from Delft) attends Delft Congress and writes report on the Congress
	• J. P. Den Hartog arrives (from Delft) in USA, begins working with Timoshenko
	• Hunsaker attends Delft Congress as stand-in for Ames
1926	• 2nd International Congress held in Sept. 12–17 in Zurich, Switzerland
	• von Kármán and Timoshenko presented papers at Zurich Congress
	• U.S. attendees include: G. Beggs (Princeton), P. W. Bridgmann (Harvard), J. A. van den Brock (Michigan)
	• Struik arrives in the U.S. from Delft and begins work at MIT where he stays until retirement
	• von Kármán visits Caltech
1927	• Timoshenko accepts faculty position at the University of Michigan
1930	• 3rd International Congress held in Stockholm, Sweden, Aug. 24–29
	• von Kármán, Timoshenko, and Dryden present papers in Stockholm
	• von Mises and den Hartog present papers in Stockholm
	• von Kármán accepts position as Director of the Guggenheim Aeronautical Laboratory at the California Institute of Technology (GALCIT) and of the Guggenheim Airship Institute at Akron, Ohio
	• J. S. Ames (Hopkins) and J. C. Hunsaker (U.S. Navy) are on International Congress Committee
	• von Mises (Berlin) on International Congress Committee
1934	• 4th International Congress of Applied Mechanics at Cambridge, UK
	• von Kármán, Dryden, and den Hartog present papers at Cambridge Congress
	• von Kármán and Timoshenko are members of the International Congress Committee
1936	• Timoshenko moves from Michigan to Stanford
1938	• 5th International Congress of Applied Mechanics in Cambridge, USA
	• von Kármán, Timoshenko, and Dryden present papers in Cambridge Congress
1939	• Richard von Mises (from Austria by way of Turkey) accepts position at Harvard
1946	• 6th International Congress of Applied Mechanics in Paris, France
	• IUTAM formally proposed
	• September 26, 1946, date of formal approval of IUTAM constitution
	• von Kármán and Dryden present papers at Paris Congress
	• Timoshenko does not attend Paris Congress

(continued)

(continued)

Year	Major USA-related mechanics developments
1948	• January 28, USNC/TAM organizational meeting held in New York, attendees: S. Wilmot (ASCE), L. A. Burckmycr , Jr. (AIEE), H. L. Dryden (IAS), S. L. Tyler & A. B. Newman (AIChE), J. R. Kline (AMS), H. A. Barton (APS), and C.E. Davies, H. W. Emmons and S. E. Reimel (ASME)
	• April 8, Dryden reported that USNC/TAM was encouraged to participate in IUTAM, attendees: H. L. Dryden, Th. von Kármán, R. R. Dexter (IAS), E. Reissner (AMS), R. J. Seeger (APS), C. E. Davies (ASME), and S. E. Reimel (ASME-committee on International Relations)
	• December 2, USNC/TAM charter approved by the U.S. Committee
1949	• March 28, USNC/TAM authorizes application for admission to IUTAM
1958	• Andre C., Simonpietri , Assistant Director of International Relations of the National Academy of Sciences, suggests that USNC/TAM become a committee of the National Academy of Sciences-National Research Council (NAS-NRC)
1966	• USNC/TAM formally approved as a committee in Division of Physical Sciences, NAS-NRC

The Independent Years 1948–1966

Little detail about operations of the Committee is known for the years 1948–1976. Three resources that are available are the IUTAM annual reports posted on the IUTAM website, Naghdi's 1979 *History of the ASME Applied Mechanics Division*, and 1966 files from the National Academy of Sciences. The IUTAM annual reports provide the memberships of the Bureau and the General Assembly, with country affiliations; beginning with 1978, membership in the IUTAM Congress Committee also is provided. The 1966 NAS files provide the minutes of two 1966 USNC/TAM meetings, actions of the NAS Governing Board approving USNC/TAM as a Committee of the Division of Physical Sciences, and the USNC/TAM constitution as of September 1971.

At a March 28, 1949 meeting in New York, the Committee authorized the application for admission to IUTAM. In a letter dated September 3, 1949, J. M. Burgers (IUTAM secretary) wrote that USNC/TAM was admitted as an adhering body to IUTAM. Thus, 1949 was the official year that USNC/TAM became a member of IUTAM.

The available USNC/TAM minutes are for the two 1966 meetings, the 1976 meeting, and from 1979 forward. The minutes are available largely due to the policies initiated by Phil Hodge when he was secretary of the Committee. Attempts to retrieve the missing minutes from former members, ASME headquarters, and the NAS were unsuccessful. It appears that for ASME and the NAS, buildings that might have housed such minutes suffered major damage with the result that records are not available.

During the early years of the Committee, many of the prime movers also were associated closely with the Applied Mechanics Division (AMD) of ASME. Timoshenko, den Hartog, Dryden, Emmons, Hoff, Hetenyi, Prager, Reissner,

Drucker, Carrier, Hodge, and Crandall all served as Chair of the AMD between 1941 and 1969. This close association between AMD and the USNC/TAM resulted in assistance from ASME for meeting space and secretarial assistance. The office of the USNC/TAM was located at ASME headquarters in New York.

C. E. Davies (ASME Secretary, the chief administrative officer of ASME from 1934 to 1957) served as Secretary of USNC/TAM for the 10 years 1948–1958. O. B. Schier II (ASME Executive Director and Secretary-Treasurer from 1957 to 1972) served as Secretary of the USNC/TAM for the 12 years 1958–1970. Thus, the chief executive officer of ASME was the Secretary of USNC/TAM from 1956 to 1970.

During the period 1948–1970, the Chairs of USNC/TAM (and their association with ASME-AMD) were: Hugh Dryden (1948–1956; AMD Chair 1942), Nick Hoff (1956–1960; AMD Chair 1955), William Prager (1960–1962; AMD Chair 1959), Miklos Hetenyi (1962–1964; AMD Chair 1957), François Frenkiel (1964–1966), Dan Drucker (1966–1968; AMD Chair 1964) and George Carrier (1968–1970; AMD Chair 1965). Thus, all of these men, except Frenkiel, served as Chair of the ASME Applied Mechanics Division prior to their assuming the leadership of USNC/TAM. Frenkiel served as Chair of the Division of Fluid Dynamics of the American Physical Society.

Timoshenko and von Kármán played major roles in the development of mechanics as an engineering science discipline in the United States. Timoshenko was the prime mover in the establishment of the ASME Applied Mechanics Division. von Kármán played a key role in the establishment of the mechanics discipline at Caltech. However, the only evidence that either was involved with the USNC/TAM is through their membership on the Committee because they were members of the IUTAM General Assembly. For Timoshenko, this is consistent with his apparent aversion for administrative duties.

A Committee of the National Academy of Sciences

In April 1958, A. C. Simonpietri (Assistant Director of International Relations of the National Academy of Sciences), brought to the attention of N. J. Hoff (Chairman of UNSC/TAM) the advantages of being represented in the international field by the National Academy of Sciences. At that time, all U.S. National Committees participating in the International Unions affiliated with the International Council of Scientific Unions, except USNC/TAM, were committees of the National Academy of Sciences-National Research Council (as it was then referred to). Serious consideration was given to this suggestion by the USNC/TAM, but no action was taken until 7 years later in 1965.

After extensive discussions between F. N. Frenkiel and D. C. Drucker (the then current Chair and Vice-Chair of USNC/TAM) with E. C. Rowans of the Foreign Secretary's Office of the National Academy of Sciences, the charter of USNC/TAM was amended on July 15, 1965. Unfortunately, the substance of those discussions is unknown.

USNC/TAM officially became a U.S. National Committee of the Division of Physical Sciences, National Academy of Sciences when the NAS Governing Board approved the USNC/TAM Constitution and by-laws on April 24, 1966. Thus, for the initial 18 years of USNC/TAM (1948–1966), the committee acted as a self-governing organization independent of any government affiliation. For the final 17 of these years it was also the U.S. adhering body of IUTAM.

Summary Minutes of the Governing Board of the NAS-NRC show that on June 6, 1965, the Board "voted to approve the proposal of the U.S. National Committee for the International Union of Theoretical and Applied Mechanics that adherence to the Union on behalf of the United States be effected through the Academy Research Council and for that purpose the committee become a committee of the Division of Physical Sciences, these steps to be effective upon approval by the Union of this manner of adherence, and approval by the Governing Board of a constitution for the U.S. National Committee."

The following letter, dated July 20, 1966, from Harrison Brown, the Foreign Secretary of the NAS was sent to Joseph J. Sisco, Assistant Secretary of State for International Affairs in the Department of State, requesting that the International Union of Theoretical and Applied Mechanics (IUTAM) be designated as an "Associated Union" of the International Council of Scientific Unions (ICSU) for which dues may be paid by the Department of State.

Dear Mr. Sisco:

The purpose of this letter is to request that the Secretary of State designate the International Union of Theoretical and Applied Mechanics (IUTAM) as an "Associated Union" of the International Council of Scientific Unions (ICSU) for which dues may be paid by the Department of State as provided under Public Law No. Z53 of the 74th Congress, SJ Resolution No. 85 of the 85th Congress and H. R. 8862 of the 89th Congress.

Since its establishment in 1947, the International Union of Theoretical and Applied Mechanics has been one of the member unions of the Council. Until recently, the U.S. National Committee for the IUTAM had been outside the framework of the National Academy of Sciences—National Research Council. However, as anticipated last year, the U.S. National Committee for the IUTAM has become a committee functioning under the aegis of the Academy Research Council, as of this year. For the information of the Department, the following IUTAM documents were sent to the Office of International Administration on May 27, 1966: Financial Report for 1965, Statutes of the Union, a list of Adhering Organizations and a chart showing the categories of adherence, The U.S. adherence is in the top category, the annual membership dues to IUTAM are $840 (12 contribution units of $70).

The General Assembly of the Union met on June 25, 1966 in Vienna, Austria and generally discussed an increase. However, final decision on the matter has been postponed to the next General Assembly, 2 years from now, when an increase in the range of 50 per cent can be anticipated.

As in the case of other international scientific unions in which the Academy Research Council participates with the cooperation of the Department, we will be pleased to keep the Department advised of major developments within the International Union of Theoretical and Applied Mechanics.

Sincerely yours,
Harrison Brown
Foreign Secretary

USNC/TAM Constitution and By-Laws

The constitution of the USNC/TAM, as of 2012, is presented in Appendix B and the by-laws are in Appendix C. The constitution of the USNC/TAM includes the following rather simple statement of purpose for the Committee:

The U.S. National Committee on Theoretical and Applied Mechanics:

(a) Will promote theoretical and applied mechanics in the United States.
(b) Will strive to maintain a balance between the various established subfields of mechanics and to accommodate and include emerging subfields.
(c) Will represent the United States in the International Union of Theoretical and Applied Mechanics on behalf of the National Academy of Sciences.

Following is a listing of modifications since the original constitution was adopted in 1958.

- 1958—Adopted by the Committee, June 13, 1958.
- 1959—Approved by participating Societies, March 25, 1959.
- 1965—Amended, July 15, 1965.
- 1966—Ratified by the Governing Board of the National Research Council, April 24, 1966.
- 1969 Constitution amended to include members of IUTAM Congress Committee as members of USNC/TAM.
- 1971—Constitution amended to add SIAM as a member society, and to increase the number of members-at-large from 5 to 7.
- 1980—Amended to change the non-voting ex-official members to be the Foreign Secretary of the NAS, one designated by the Assembly of engineering, two designated by the Assembly of Mathematical and Physical Sciences.
- 1981—Amended to add SNAME as a member society.
- 1982—Amended to add SES as a member society.
- 1985—Amended because SESA names changed to SEM.
- 1993—Amended to add AAM as a member society.
- 1992–1993—changed to by-laws about new member societies.
- 1993—Amended to add ASA as a member society.
- 2001—A small expansion of the Purpose of the Committee was approved.
- 2006—Amended to add USACM as a member society.
- 2008—Amended to change Member-at-Large terms from 2 to 3 years.
- 2012—Amended to change status of IUTAM Members-at-Large to Honorary non-voting members of USNC/TAM.

USNC/TAM Member Societies

The USNC/TAM started with seven member societies in 1948. As the following table indicates, additional societies were added at intermittent intervals until the total number reached 15 in 2006. It is interesting that societies were added in intervals of roughly 10 years. It is fair to say that most societies with major activities in mechanics are represented in USNC/TAM. The constitution lists the member societies. As a two-thirds vote is required to change the constitution, this means that new society members can be admitted only upon approval of two-thirds of the voting members. Information on each of the member societies is presented in Appendix D.

| USNC/TAM member societies by year | | | | | | | | |
Yr/No.	7	8	9	10	11	12	13	14	15
1948–1958	AIAA, AMS, APS, ASME, SESA, ASCE, AIChE								
1960		SOR							
1961			ASTM						
1971				SIAM					
1981					SNAME				
1982						SES			
1992							AAM		
1993								ASA	
2006									USACM

USNC/TAM Members

Unfortunately, membership records for the early years of the USNC/TAM are incomplete. Annual Reports for the entire history of IUTAM (made available in May 2014) provide information on the U.S. Representatives and Elected Members of the IUTAM General Assembly. As specified in the USNC/TAM constitution, these individuals are members of the USNC/TAM. However, it is generally true that while we know that these individuals are members of the USNC/TAM, we do not know if they were Society Representatives (for 1949–1958), Members-at-Large (for 1949–1969) or serving in IUTAM in another capacity such as a member of the IUTAM Congress Committee.

   Elected Members of IUTAM (normally) serve 4-year terms. These *Elected Members* (also called Personal Members) are IUTAM *Member-at-Large*. In the early years of IUTAM, *Elected Members* were automatically re-elected every 4 years throughout their life. They were *member for life*. This is no longer the case and re-election is no longer automatic. For the members elected for life, the year of their death is noted in the membership roster. As of this writing, Bruno Boley is the only remaining U.S. person who is an IUTAM *member for life*.

Starting in 1965, the IUTAM Annual Reports provide information on the Members of the IUTAM Congress Committee. As per the USNC/TAM constitution, U.S. members of the IUTAM Congress Committee are members of the USNC/TAM.

Using data from the IUTAM Annual Reports, a partial listing of the USNC/TAM membership for the early years of the Committee was assembled. This list includes only those members who had official appointments in IUTAM.

Naghdi's report provides the committee membership for 1948 and it is assumed that the membership was the same for 1949. For the period 1950–1982, the IUTAM Annual Reports and data assembled by Hodge provide what is known about the membership. From 1982 forward, minutes of USNC/TAM meetings provide the committee membership. Listing the membership at the time of meetings clearly indicates those who had responsibility for the decisions made.

We note that for both the USNC/TAM and IUTAM, terms of appointment start on November 1 and end on October 31; the length of the term is variable depending upon the type of appointment.

Based upon the data available there have been at least 190 voting members of the USNC/TAM through 2012. Ex-officio, non-voting members have been appointed to the committee by the NAS/NRC. With the exception of representatives from the materials community who regularly attended meetings, no attempt has been made to include these individuals in the membership listing. Alternates have attended meetings at various times, and on occasion, substitute secretaries took minutes. No attempt has been made to record these occurrences.

Appendix E provides listings of the membership by category, committee membership by year (Appendix E.1), officers (Appendix E.2), society representatives (Appendix E.3), members-at-large (Appendix E.4) and U.S. congress chairs (Appendix E.5).

Meeting dates of the Committee are provided when known. Meetings in Washington DC typically were held on Friday and Saturday. Meetings held in conjunction with a U.S. Congress (every 4 years) were held at the Congress site on Sunday prior to the Congress.

USNC/TAM Officers

The number and type of committee officers has varied over the years. A copy of the constitution and by-laws dated Sept. 25, 1985 is in the minutes of the May 1989 meeting. This 1985 constitution states that there are to be three officers: Chairman, Vice-Chairman, and Secretary. It also states that the two most recent retired Committee Chairman and Personal Members (also referred to as IUTAM Elected Members) of the IUTAM General Assembly are to be members of the USNC/TAM. Later changes to the constitution fixed the number of officers as four: Chair, Vice-Chair, Secretary, and Past Chair.

Chapter 10
USNC/TAM Chairs

Twenty-eight individuals served as the Chair of USNC/TAM between 1948 and 2012. All served one 2-year term except for the first two Chairs of the Committee, Hugh Dryden served 8 years, 1948–1956, and Nick Hoff served 4 years, 1956–1960. Following are brief highlights of the committee chairs. It is evident that the Chairs of USNC/TAM have outstanding records of accomplishment and are well known to the mechanics community. Major accomplishments, awards, and positions of the Chairs are presented. However, detailed lists of publications and society fellow memberships, are not provided as the length of such lists would be exhaustive. There can be no question that in future years, many of these chairs will be considered *giants of mechanics*. I have had the good fortune of personally knowing all but a small handful of the committee chairs.

Hugh L. Dryden (Chair: 1948–1956)

In 1934, Dryden was appointed Chief of the Mechanics and Sound Division at NBS. In 1939 he became a member of the National Advisory Committee for Aeronautics (NACA was the forerunner of NASA). He headed a group at the Bureau of Ordnance Experimental Unit (within the National Bureau of Standards) which developed the radar homing missile, BAT. He also served as Deputy Director of the Army Air Force's Scientific Advisory Group (headed by von Kármán) during the war years. Dryden served as the Director of NACA from 1947 to 1958. When NACA became NASA, he served as Deputy Director of NASA until his

© Springer International Publishing Switzerland 2016
C.T. Herakovich, *Mechanics IUTAM USNC/TAM*,
DOI 10.1007/978-3-319-32312-1_10

death in 1965. He was editor of the Journal of the Institute of the Aeronautical
Sciences from 1941 to 1956. Dryden was Chair of the ASME Applied Mechanics
Division in 1942.

As discussed in a previous Chapter, Dryden played a key role in the establishment of the USNC/TAM and served as its Chair for the initial 8 years of the
committee. He is the only person to have served as Chair for 8 years.

Dryden was a founding member of the National Academy of Engineering and a
member of the National Academy of Sciences. The NASA Dryden Flight Research
Center in California is named after him as is the Dryden crater on the Moon. He
was honored with many awards including the National Medal of Science in
Engineering in 1965. Other awards include: Wright Brothers Lecture of the
Institute of the Aeronautical Sciences (1938), U.S. Medal of Freedom (1946),
Sylvanus Albert Reed Award of the I.A.S. (1940), Order of the British Empire
(civil division) (1948), Presidential Certificate of Merit (1948), 37th Wilbur Wright
Memorial Lecture of the Royal Aeronautical Society (1949), Daniel Guggenheim
Medal (1950), Wright Brothers Memorial Trophy (1955), Ludwig Prandtl
Memorial Lecture of the Wissenschaftliche Gesellschaft fur Luftfahrt (1958),
Elliott Cresson Medal of the Franklin Institute (1961), First von Kármán Lecture,
American Rocket Society (1962), Goddard Memorial Trophy (1964), ASME
Thurston Lecture (1965), ASME Honorary Member (1964). He received Honorary
Degrees from 15 universities.

His research on the problems of wind tunnel turbulence and boundary layer flow
resulted in international recognition. Dryden's leadership in the development of
high-speed wind tunnels, flight testing, and a competence for theoretical research
within NACA contributed substantially to the leadership of the United States in
supersonic flight.

Dryden was elected to the National Academy of Sciences in 1944 and served as
Chairman of its Section on Engineering from 1947 to 1950. In 1955 he was elected
Home Secretary, a position he held until his death in 1965.

Together with von Kármán, he was an editor of *Applied Mechanics Reviews*.
Contributions by Dryden may be found in the Proceedings of the 3rd, 4th, and 5th
International Congresses of Applied Mechanics. He contributed to *Advances in*

Applied Mechanics, Vol. I, 1948, with authorship of the section entitled *"Recent Advances in the Mechanics of Boundary Layer Flow."*

Dryden gave the 1965 ASME Thurston Lecture 1 month before his death at the age of 67.

Nicholas J. Hoff (Chair: 1956–1960)

Nick Hoff was born in Magyarovar, Hungary in 1906. In 1928, he received his Dipl. Ing. from the Swiss Federal Institute of Technology in Zurich. From 1929 to 1938 he worked as an aircraft stress analyst and designer in Budapest. He arrived at Stanford in 1938 to study under Stephen Timoshenko, receiving his Ph.D. in 1942. He joined the Polytechnic Institute of Brooklyn in 1940 as instructor in aeronautical engineering, becoming full professor in 1946 and Head of the Department of Aeronautical Engineering and Applied Mechanics in 1950. Upon the recommendation of von Kármán who was at Caltech, Hoff was appointed Head of the new Department of Aeronautics at Stanford in 1957. In 1962, the department name was changed to Aeronautics and Astronautics.

Hoff (two terms) and Dryden (four terms) are the only two individuals to have served more than one term as Chair of the USNC/TAM. Hoff became a U.S. Representative to the IUTAM General Assembly in 1953 and a member of the IUTAM Bureau in 1961. He continued as a member of the Bureau until 1971.

Hoff's work on stability of thin-walled structures resulted in many honors and awards in the U.S. and abroad. He was Chair of the ASME Applied Mechanics Division in 1955. His honors include: ASME Worcester Reed Warner Medal (1967), ASCE von Kármán Medal (1972), ASME Medal (1974), and ASME Honorary Member (1983), Daniel Guggenheim Medal (1983), AIAA Structures, Structural Dynamics, & Materials Award (1971), and AAM Outstanding Service Award (1990). Hoff was elected a member of the U.S. National Academy of Engineering in 1997.

Nick Hoff died on August 4, 1997.

William Prager (Chair: 1960–1962)

William Prager was born on May 23, 1903 in Karlsruhe, Germany. He received his Dipl. Ing. Degree in 1925 and his doctorate in engineering in 1926, both from the Technical University of Darmstadt. At the age of 26 in 1929, he was appointed Director of the Institute of Applied Mechanics at Göttingen, and in 1932 was appointed Professor at the Institute of Technology at Karlsruhe. Forced out of his professorship at Karlsruhe for his anti-Nazi views in 1934, he sued the German government and won a year of back pay and an offer to return to his position. He felt it was best to leave Germany and accepted an offer as Professor of Mechanics at Istanbul University.

With the outbreak of the Second World War, he decided to immigrate to the U.S. to accept the new position of Director of Advanced Instruction and Research in Mechanics at Brown University. It took him, his wife, and 12-year-old son 40 days to reach the U.S. by a circuitous route. They arrived at Brown in November 1941. Prager founded the Quarterly of Applied Mathematics in 1943 and edited it for over 20 years. He established the Division of Applied Mathematics at Brown in 1946 and served as its first Chairman. He is best known for his work in continuum mechanics and plasticity. His books have been published in several languages.

Prager received many honors including membership in the National Academy of Engineering (1965), the National Academy of Sciences (1968), the American Academy of Arts and Sciences, the Polish Academy of Sciences, and the French Académie des Sciences. He received the ASME Worcester Reed Warner Medal (1957), ASME Honorary Member (1970), the ASME Timoshenko Medal (1966), and the first ASCE von Kármán medal (1960). Many universities awarded him honorary degrees including Liege, Poitiers, Milan, Waterloo, Stuttgart, Hannover, Brown, Manchester, and Brussels. SES named the Prager Medal in his honor (1983). He was Chair of the ASME Applied Mechanics Division in 1959. Prager died on March 16, 1980 in Zurich, Switzerland.

Miklos Hetenyi (Chair: 1962–1964)

Miklos Hetenyi was born in Debrecen, Hungary on November 5, 1906. He earned his diploma in civil engineering from the Budapest University of Technical Sciences in 1931. He came to the United States in 1934 to study with Timoshenko at the University of Michigan, receiving his Ph.D. in Engineering Mechanics in 1936. After working at Westinghouse for 12 years, he became professor at Northwestern and then transferred to Stanford's Applied Mechanics Department in 1962.

It is said that he was one of Timoshenko's favorite students, competent in the mathematical theory of elasticity and experimental stress analysis. He worked closely with Ray Mindlin and Dan Drucker on experimental work. He was well known for his 1946 book Beams on Elastic Foundations and as Editor-in-Chief of the Handbook of Experimental Stress Analysis. He is credited as one of the founders (with William Murray and Raymond Mindlin) of the Society of Experimental Stress Analysis (now Society of Experimental Mechanics).

Hetenyi received many honors from societies and universities including the Miklos Hetenyi Award named in his honor by the Society of Experimental Mechanics. He was Chair of the ASME Applied Mechanics Division in 1957 and Honorary Member of ASME (1973).

Hetenyi died on Oct. 31, 1984, at his desk in his home on the Stanford campus while working on the manuscript of a book on structural analysis.

François N. Frenkiel (Chair: 1964–1966)

François Naftali Frenkiel, was born in Warsaw, Poland, on September 19, 1910. He received his undergraduate education in Mechanical and Aeronautical Engineering at the University of Ghent, Belgium, and his Ph.D. in Physics from the University of Lille in France. Following his Ph.D., he worked at the French Aeronautical Research Station at Toulouse. He was imprisoned in concentration camps by the Nazis for 2 years. His pregnant first wife died in one of the camps. Frenkiel was freed from Buchenwald prison in 1945 by the Allied armies.

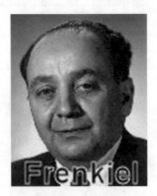

Frenkiel arrived in the United States in 1947 (von Kármán was his sponsor) and over time was associated with Cornell University, the U.S. Naval Ordnance Laboratory, the Johns Hopkins University Applied Physics Laboratory, and from 1960 until his retirement with the David W. Taylor Naval Ship Research and Development Center.

Frenkiel was the founder (1958) and longtime editor (until 1981) of the *Physics of Fluids*. He was the chair and secretary of the APS Division of Fluid Dynamics on numerous occasions during the period 1954–1970. The APS, *François Frenkiel Award for Fluid Mechanics* is named in his honor.

He served with the International Union of Theoretical and Applied Mechanics in several capacities. He was a member of the fluids symposium panel (1979–1984) and an elected Member-at-Large (1980–1988). Frenkiel was a member of the USNC/TAM for 29 years, 1956–1985. In addition to being chair of the committee for 1964–1966, he served as Secretary of USNC/TAM for the 12 years 1970–1982.

Frenkiel published extensively in the field of turbulent flow and pioneered the application of high-speed digital computing methods to the measurement of turbulence and the mathematical modeling of urban pollution.

Frenkiel died in Rockville, MD at 75 on July 9, 1986.

Daniel C. Drucker (Chair: 1966–1968)

Dan Drucker was born on June 3, 1918 in New York City. He received D.S., M.S. and Ph.D. degrees from Columbia University where his 1940 Ph.D. was on photo-elasticity under Ray Mindlin. Drucker taught at Cornell from 1940 to 1943. Then periods of work were conducted at the IIT Armour Research Foundation, the U.S.

Army Air Corps, and back to the Illinois Institute of Technology before he went to Brown University in 1947. At Brown, he did pioneering work on plasticity where he often worked closely with Prager. Drucker may be best known for his stability postulate for plasticity theory.

In addition to his outstanding research contributions, Drucker served the engineering profession with dedication and distinction. He was president of AAM, ASME, ASEE, SESA, and IUTAM. He was Dean of Engineering at Illinois for 15 years beginning in 1968. He served as Editor of the ASME Journal of Applied Mechanics for 12 years. Drucker became a U.S. Representative to the IUTAM General Assembly in 1964, IUTAM Treasurer 1972–1980, IUTAM President 1980–1984, and IUTAM Vice-President 1984–1988. He was a member of the USNC/TAM for 39 years, 1962–2001, and a member of the IUTAM General Assembly from for 38 years, 1964–2001. Dan was Chair of the ASME Applied Mechanics Division in 1964.

In 1997, I had the privilege, as Chair of the ASME Applied Mechanics Division, of writing to Drucker to inform him that the ASME Daniel C. Drucker Medal had been established in his honor and that he was to be the first honoree in 1998. He was recognized with numerous other awards, including: the ASME Timoshenko Medal (1983), the ASME Medal (1992), ASME Honorary Membership (1982), ASME Thurston Lecture (1986), National Medal of Science (1988), the first SES Prager Medal (1983), ASCE von Kármán Medal (1966), SEM Honorary Membership, William M. Murray Lecture, and M. M. Frocht Award, AAM Outstanding Service Award (1986), University of Liege Gustav Trasenter Medal, Columbia University Egleston Medal and the Illig Medal, Founder Engineering Societies John Fritz Medal, ASTM Marburg Lecture (1966), the Prof. Modesto Panetti and Prof. Carlo Ferrari International Prize and Gold Medal.

Honorary doctorates were conferred by Lehigh, Technion, Brown, Northwestern, and the University of Illinois at Urbana-Champaign. He was a member of the National Academy of Engineering (1967), the American Academy of Arts and Sciences, and a Foreign Member of the Polish Academy of Sciences. He completed his academic career at the University of Florida.

Dan Drucker died in Gainesville, Florida on September 1, 2001.

George F. Carrier (Chair: 1968–1970)

George Francis Carrier was born on May 4, 1918 in Millinocket, Maine. He attended Cornell University and received his B.S. in Mechanical Engineering in 1939 and Ph.D. in 1944. His advisor at Cornell was J. N. Goodier. Carrier was an outstanding mathematician. He was particularly noted for his ability to intuitively model a physical system and then deduce an analytical solution. While his Ph.D. thesis dealt with aeolotropic (anisotropic) solids, his later work was primarily on fluid mechanics, combustion, and tsunamis. After his Ph.D., he worked with Howard Emmons at Harvard for 2 years on compressible fluids including helping to design, build, and work in a high-speed wind tunnel. He joined the faculty at Brown as an assistant professor in 1946 and was a full professor by 1947. In 1952, he returned to Harvard as Gordon McKay Professor; he stayed at Harvard for the remainder of his career.

Carrier was a member of the National Academy of Sciences (1967), the National Academy of Engineering (1974), the American Academy of Arts and Sciences, and the American Philosophical Society (1976). He was Chair of the ASME Applied Mechanics Division in 1965. He received awards from many professional societies including SIAM's von Neumann Lecturer (1969) and von Kármán Prize (1979), ASME's Timoshenko Medal (1978), Thurston Lecture (1972), and Russ Richards Award (1963), ASCE von Kármán Medal (1977). He received the National Medal of Science (1990).

E. H. Lee (Chair: 1970–1972)

Erastus (Ras) H. Lee was born on February 2, 1916, in Southport, England. He earned a bachelor degree in mechanical sciences and mathematics from Cambridge University in 1937. Following a year of postgraduate study at Cambridge, he went to Stanford University to study with Timoshenko. He completed the Ph.D. at Stanford in 1940 and then worked in the British war effort, initially in New York City and Washington, DC, and then back in England. By 1946, he was an Assistant Director in the British Department of Atomic Energy. An invitation from Prager brought Lee to Brown University in 1948 where he stayed for 14 years as Professor of Applied Mathematics. He was Chairman of the Applied Mathematics Department at Brown for 5 years. While at Brown, Lee worked alongside Prager, Drucker, Kolsky, Rivlin, Shield and Sternberg, among others.

In 1962, Lee accepted appointment as Professor in the Division of Applied Mechanics in the Department of Aeronautics and Astronautics at Stanford. At Stanford, Lee worked with Goodier, Flugge, Hoff and Hetenyi. Upon mandatory retirement at age 65 from Stanford, Lee spent 10 more years as Professor of Engineering at Rensselaer Polytechnic Institute.

Lee made fundamental contributions in plasticity, viscoelasticity, and wave propagation in solids. He was a member of the National Academy of Engineering (1975) and received the ASME Timoshenko Medal (1976). Lee was a member of the IUTAM General Assembly from 1969 to 1975, and a member of the IUTAM Bureau 1972–1974.

Ras Lee died at the age of 90 on May 17, 2006 in Lee, New Hampshire.

Stephen H. Crandall (Chair: 1972–1974)

Stephen Harry Crandall was born in the Philippine Islands on December 2, 1920. Following a degree in mechanical engineering at Stevens Institute of Technology in Hoboken, New Jersey, in the early days of WWII, Crandall joined the MIT Radiation Laboratory as a staff member. He earned a Ph.D. in mathematics from MIT (1946) where he worked under J. P. Den Hartog. He joined the MIT faculty in 1946 and taught dynamics and strength of materials until his retirement in 1991. He was a prolific author of texts in solid mechanics, numerical methods, and random vibrations. Crandall was Chair of the ASME Applied Mechanics Division in 1969.

ASME recognized his work with the Worcester Reed Warner Medal (1971), Honorary Membership (1988), Thurston Lecture (1988), Timoshenko Medal (1990), Den Hartog Medal (1991), and Thomas K. Caughey Dynamics Award (2009). He received ASA's Trent-Crede Medal (1978), ASCE von Kármán (1984) and Freudenthal (1996) Medals and AAM Outstanding Service Award (1993). Crandall was elected to the National Academy of Engineering (1977) and the National Academy of Sciences (1993). He was also elected to the American Academy of Arts and Sciences and the Russian Academy of Engineering.

Stephen Crandall passed away on Oct. 29, 2013 in Needham, MA at age 92.

Bruno A. Boley (Chair: 1974–1976)

Bruno A. Bolaffio was born in Gorizia (near Trieste), Italy, in 1924. His family immigrated to the United States in 1939 where the family name was changed to Boley. He received a B.S. in civil engineering from the College of the City of New York in 1943, a Sc.D. in aeronautical engineering from the Polytechnic Institute of Brooklyn in 1946, and an honorary Sc.D. degree from the College of the City of New York in 1982. He served as Assistant Professor, Aeronautical Engineering at the Polytechnic Institute of Brooklyn from 1943 to 1948, and worked at Goodyear Aircraft Corporation from 1948 to 1950. Boley then returned to academic life as Assoc. Professor of Aeronautical Engineering at Ohio State University. In 1952, he moved to Columbia University where he stayed from 1952 to 1968 as Prof. of Civil Engineering. Boley was professor and chair of the Department of Theoretical and Applied Mechanics Department at Cornell from 1968–72. From 1972 to 1986 Boley was Dean of Engineering at Northwestern University; following his deanship he was professor emeritus. He was visiting professor at the Technical University of Milan in 1964–1965 and at the Imperial College of Science and Technology, University of London in 1972.

Boley has been a member of the USNC/TAM continuously since 1968. He served on the Congress Committee as: Member (1968–1996), Executive Committee (1972–1982), and Secretary (1976–1983). He became a voting member of the IUTAM General Assembly in 1976 and an elected member of the General Assembly beginning in 1980 and will remain a member throughout his life.

Boley is a member of the National Academy of Engineering (1975), past president of the American Academy of Mechanics and the Society of Engineering Science. He was Chair of the ASME Applied Mechanics Division (1975), and served on the Board of Governors of ASME and the Argonne National Laboratory. ASME recognized Boley with Honorary Membership (1980), Worcester Reed Warner Medal (1991), and Drucker Medal (2001). ASCE honored Boley with the von Kármán Medal (1991), AAM with the Outstanding Service Award (1987). Boley is known for his work in structural dynamics, elastic stability, applied mathematics, thermal stresses, and heat conduction. His book *Theory of Thermal Stresses* with J. H. Weiner (1960) is considered a classic.

Ronald S. Rivlin (Chair: 1976–1978)

Ronald Rivlin was born on May 6, 1915 in London, England. He received an undergraduate degree in physics and mathematics from St. John's College, Cambridge in 1937, a master's in 1939, and a doctorate from Cambridge in 1952. Work experience included 5 years with General Electric in Wembley, England, 2 years with the Ministry of Aircraft Production (1942–1944), and the British Rubber Producers Research Association. In 1952 Rivlin spent a year as a consultant to the Naval Research Laboratory in Washington, DC; he then decided to stay on in the United States. From 1953 to 1967, he was professor (and Chair) at Brown in the Division of Applied Mathematics. In 1967, he moved to Lehigh as Director of the

Center for the Application of Mathematics. Following retirement as Director of the Center in 1980, he remained as adjunct professor for another 10 years.

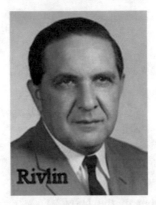

Rivlin made fundamental contributions in continuum mechanics. He is associated with theories for large elastic deformations. He developed finite strain theories for Neo-Hookean solids that exhibit nonlinear stress–strain response and his name is associated with what are called Mooney-Rivlin solids.

Rivlin's honors included the ASME Timoshenko Medal (1987), the ASCE von Kármán Medal (1993), the SOR Bingham Medal (1958), and the ACS Goodyear Medal (1992). Rivlin was a member of the National Academy of Engineering (1985), as well as the National Academy of Arts and Sciences, the Academia Nazionale dei Lincei and the Royal Irish Academy. He was Chair of the ASME Applied Mechanics Division in 1980.

Rivlin died on Oct. 4, 2005.

Paul M. Naghdi (Chair: 1978–1980)

Paul Naghdi was born in Tehran, Iran on March 29, 1924. In 1943, in order to pursue his education, he undertook a perilous ship voyage to the United States, during which he helped navigate the ship. Naghdi studied mechanical engineering at Cornell, graduating in 1946. After a short period in the U.S. Army, he enrolled in the graduate program in Engineering Mechanics at the University of Michigan. He received an M.S. degree in 1948 and his doctorate in 1951. Naghdi was granted U.S. citizenship in 1948. He taught in the Engineering Mechanics Department at Michigan from 1949 to 1958, rising through the ranks from instructor to full professor in 1954.

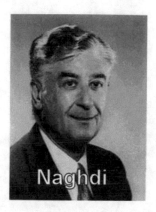

Naghdi moved to the University of California at Berkeley in 1958 as professor of Engineering Science; there he led the establishment of the Division of Applied Mechanics in the Mechanical Engineering Department. Naghdi worked primarily in the area of continuum mechanics; he also was known for his work in shell theory and plasticity. He was a very active member of the ASME mechanics community including Chair of the Applied Mechanics Division (1972) and author of a history of the Division for its 50th Anniversary. He was awarded the Timoshenko Medal (1980), made an Honorary Member of ASME (1983), and was elected to the National Academy of Engineering (1984). SES awarded him the Eringen Medal in 1986. He received Honorary Doctoral Degrees from the National University of Ireland (1987) and Université Catholique de Louvain (1992).

Naghdi's *Brief History of the Applied Mechanics Division of ASME* (1979) provided a permanent record of the contributions of many outstanding mechanicians to ASME and the Division.

Joseph B. Keller (Chair: 1980–1982)

Joe Keller was born in Paterson, New Jersey on July 31, 1923. He earned three degrees from New York University, B.A. (1943), M.S. (1946), and Ph.D. (1948). His Ph.D. advisor was Richard Courant. Keller remained at NYU eventually becoming Professor of Mathematics in the Courant Institute of Mathematical Sciences. After 30 years at NYU, he moved to Stanford University in 1979 where he was Professor of Mathematics and Mechanical Engineering. Keller took emeritus status at Stanford in 1993.

Keller is an applied mathematician. His research has been concerned with the use of mathematics to solve a variety of problems in science and engineering. Some of these problems are wave propagation in solids and oceans. His name is associated with the Einstein-Brillouin-Keller method for computing eigenvalues in quantum mechanics

Keller is a member of the National Academy of Science (1974), and a Foreign Member of the Royal Society of London. He was the Honorary Professor of Mathematical Sciences at the University of Cambridge (1989–1993). He is a recipient of the Wolf Foundation's Wolf Prize (1997), the Frederick E. Nemmers Prize in Mathematics (1996), the National Academy of Sciences Award in Applied Mathematics and Numerical Analysis (1995), the National Medal of Science (1988), the ASME Timoshenko Medal (1984), the SES Eringen Medal (1981), the SIAM von Kármán Prize (1979), the Gibbs Lecturer of the American Mathematical Society (1977), and the SIAM von Neumann Lecturer (1983). He has received honorary doctorates from New Jersey Institute of Technology, the University of Crete, Northwestern University, and the Technical University of Denmark.

James W. Dally (Chair: 1982–1984)

Jim Dally was born on August 2, 1929 in Sardis, OH. He earned a B.S. (1951) and M.S. (1953) in mechanical engineering at Carnegie Institute of Technology (now Carnegie-Mellon University). His Ph.D. is from the Illinois Institute of Technology (1958). He was an assistant professor at Cornell University (1958–1961) and Assistant Director of Research at IIT Research Institute from 1961 to 1964. He taught at IIT from 1964 to 1971, the University of Maryland from 1971 to 1979 and was Dean of Engineering at the University of Rhode Island from 1979 to 1982. Dally joined IBM as the Head of mechanical development at their Manassas laboratory in 1982. In 1984, he rejoined the University of Maryland where he served until retirement in 1997. Dally is now the Glenn L. Martin Institute Professor of Engineering (emeritus) at the University of Maryland at College Park.

Dally is the author or co-author of six books, including engineering textbooks on experimental stress analysis, engineering design, instrumentation, and the packaging of electronic systems. He is a member of the National Academy of Engineering (1984), and has served on a number of National Research Council committees. Dally received the ASME Drucker Medal (2012), AAM Outstanding Service Award (2004), and the ASEE Archie Higdon Distinguished Educator Award (2013).

Andreas Acrivos (Chair: 1984–1986)

Andy Acrivos was born on June 13, 1928 in Athens, Greece. In 1947, he took a converted troop ship to the United States to begin studies at Syracuse University, the only university to offer him a scholarship, something he needed. Acrivos received a Bachelor's degree from Syracuse in 1950. He then earned a Master's (1951) and a Ph.D. (1954) from the University of Minnesota, with all three of his degrees being in chemical engineering. Following his Ph.D., Acrivos joined the faculty at the University of California, Berkeley, where he rose through the ranks. In 1962, he moved to Stanford and then, in 1987, joined The City College of New York (CCNY) as the Albert Einstein Professor of Science and Engineering. He retired from City College in 2000 and is professor emeritus at both CCNY and Stanford.

Acrivos became a member of the USNC/TAM in 1959 as the AIChE representative. He served in that capacity for three terms until 1971. He was not in the committee from 1971 to 1980, rejoined the committee in 1980 for two terms as Member-at-Large. He was Chair, Vice-Chair, and then Past Chair from 1984 to 1990. In 2004, he returned to the committee when IUTAM elected Acrivos as a Member-at-Large, a capacity he continued in through 2016.

Acrivos is a leading Chemical Engineering educator and fluid dynamicist. His awards include the National Medal of Science (2001), APS Fluid Dynamics Prize (1991), SES G. I. Taylor Medal (1988), and SOR Bingham Medal (1994). AIChE has honored Acrivos with the Allan Colburn Award (1963), Professional Progress Award (1968), and Warren K. Lewis Award (1984). He is a member of the National Academy of Engineering (1977), National Academy of Sciences (1991), and is a Fellow of the American Academy of Arts and Sciences (1993). Acrivos served as the Editor of the Physics of Fluids from 1982 to 1997. In 2014, the AIChE Professional Progress Award was renamed the AIChE Andreas Acrivos Award.

H. Norman Abramson (Chair: 1986–1988)

Norm Abramson was born in San Antonio, TX on March 4, 1926. He completed a B.S. (1950) and M.S. (1952) in Mechanical Engineering at Stanford. His Ph.D. was in Engineering Mechanics from the University Texas at Austin in 1956. Abramson was an Associate Professor at Texas A & M from 1952 to 1956. He then joined the Southwest Research Institute in March, 1956, where he eventually rose to Executive Vice-President. Abramson has served on numerous government and corporate advisory boards, and as Adjunct Professor at UT-Austin and UT-San Antonio.

Abramson was inducted into the National Academy of Engineering (1976) and has served on the NAE Council. He has been the Vice-President and governor of the American Society of Mechanical Engineers (ASME), and Director of the American Institute of Aeronautics and Astronautics (AIAA).

His awards include ASME Honorary Member (1979), ASME Medal (1999), and ASME Ted Belytschko Applied Mechanics Award (1988). He also received the AIAA Distinguished Service Award (1973), AIAA Structures, Structural Dynamics, Materials Award (1991), and the AAM Distinguished Service (1991). He works in fluid dynamics and transportation. Abramson is now the retired Executive Vice-President of Southwest Research Institute. He was Chair of the ASME Applied Mechanics Division in 1970.

Richard M. Christensen (Chair: 1988–1990)

Dick Christensen was born on July 3, 1932 in Idaho Falls, Idaho. His degrees are all in Civil Engineering, B.S. (Utah 1955), and M. Eng. (1956) and D. Eng. (1961) from Yale. From 1956 to 1958, he was at Convair Division of General Dynamics. From 1961 to 1967, he was at Space Technology Laboratories and at U. C. Berkeley. The following 7 years were with Shell Development (Shell Oil), and thereafter with Washington University. In 1976, he joined Lawrence Livermore National Laboratory as Senior Scientist in the Chemistry and Materials Science Department. In 1993, Christensen joined Stanford University as Research Professor in the Aeronautics and Astronautics Department where he is emeritus professor.

Christensen is a member of the National Academy of Engineering (1987), and Honorary Member of ASME (1992). His work has been recognized by ASME with the Worcester Reed Warner Gold Medal (1988), the Nadai Medal (2006), and Timoshenko Medal (2013). He also received the SES Prager Medal (1988). His research is concerned with mechanics of materials with emphasis on fiber composites, micromechanics, and failure criteria. He is the author of three books: *Theory of Viscoelasticity (1971)*, *Mechanics of Composite Materials (1979)*, and *The Theory of Materials Failure* (2013). Christensen was Chair of the ASME Applied Mechanics Division in 1981.

Sidney Leibovich (Chair: 1990–1992)

Sid Leibovich was born in Memphis. TN on April 2, 1939. He earned a B.S. in engineering from Caltech in 1961 and a Ph.D. in Theoretical and Applied Mechanics from Cornell in 1965. He joined the Cornell faculty after a year as a NATO postdoctoral fellow in mathematics at the University of London. Leibovich has spent his entire career at Cornell where he is now a chaired professor emeritus of Mechanical and Aerospace Engineering. At Cornell, he was the S.C. Thomas Sze Director of the Sibley School of Mechanical and Aerospace Engineering for 7 years, Director of the Center for Applied Mathematics, and founding Associate Director for Energy of the Atkinson Center for a Sustainable Future.

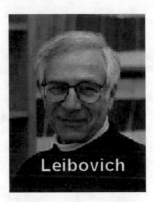

Leibovich is a member of the National Academy of Engineering (1993), and is currently the Vice-Chair of the Mechanical Engineering Section, and a Fellow of the American Academy of Arts and Sciences (1992). He previously served as general editor of the Cambridge University Press Monographs on Mechanics; co-editor of *Acta Mechanica*; and associate editor for *the Journal of Fluid Mechanics*, *the Journal of Applied Mechanics*, *the SIAM Journal on Applied Mathematics*, and as a member of the editorial board of *the Annual Review of Fluid Mechanics*. He has been Chair of the ASME Applied Mechanics Division (1989), and Chair of the APS Division of Fluid Dynamics.

Leibovich's work is concerned with fluid mechanics with emphasis on vortex flows, hydrodynamic stability, and nonlinear wave propagation

J. Tinsley Oden (Chair: 1992–1994)

Tinsley Oden was born on December 25, 1936 in Alexandria, Louisiana. He earned a B.S. (1959) and M.S. (1960) in Civil Engineering from Louisiana State University, and a Ph.D. in Engineering Mechanics from Oklahoma State University (1962). Oden taught at OSU and then The University of Alabama in Huntsville until 1973 when he joined the faculty at the University of Texas at Austin. Oden is Associate Vice-President for Research, Director of the Institute for Computational Engineering and Sciences, Cockrell Family Regents' Chair in Engineering No. 2, Peter O'Donnell Jr. Centennial Chair in Computer Systems, professor of Aerospace Engineering and Engineering Mechanics, Professor of Mathematics, and Professor of Computer Science.

Oden is a member of the U.S. National Academy of Engineering (1988), a Fellow of the American Academy of Arts and Sciences (2008), and an Honorary Member of the American Society of Mechanical Engineers (2004). He is a founding member and the first President of the U.S. Association for Computational Mechanics (USACM) and the International Association for Computational Mechanics. He is also the past President of the American Academy of Mechanics and the Society of Engineering Science. His awards include the SES A. C. Eringen Medal (1989), ASME Stephen P. Timoshenko (1996) and Worcester Reed Warner (1990) Medals, OSU Lohmann Medal (1991), ASCE Theodore von Kármán Medal (1992), JSME Computational Mechanics Award (1993), USACM John von Neumann medal (1995), AAM Outstanding Service Award (1995), and the IACM Newton-Gauss (Congress) Medal (1994). Oden was knighted as "Chevalier des Palmes Academiques" by the French government (1990). He received the SIAM Distinguished Service Award (2009) and the SIAM/ACM Prize in Computational Science and Engineering (2011).

The USACM established the J. Tinsley Oden Medal to recognize "outstanding and sustained contribution to computational science, engineering, and mathematics" in 2012. Most recently, he was awarded the 2013 Honda Prize Laureate from the Honda Foundation in Japan for his contributions in establishing the field of Computational Mechanics.

Oden holds honorary doctorates, from universities in Portugal (Technical University of Lisbon, 1987), Belgium (Faculte Polytechnique de Mons, 2000), Poland (Cracow University of Technology, 2001), France (École Normale Superieure Cachan, 2006), and the United States (The University of Texas at Austin, 2004, and Ohio State University, 2010). Oden is an Editor of Computer Methods in Applied Mechanics and Engineering, and the series, Finite Elements in Flow Problems and Computational Methods in Nonlinear Mechanics.

Tinsley Oden has made significant contributions in the mathematical theory and implementation of numerical methods applied to problems in solid and fluid mechanics and, particularly, nonlinear continuum mechanics.

Tinsley Oden is the author or co-author of 27 books, a sampling follows:

- Oden, J. T., *An Introduction to Mathematical Modeling: A Course in Mechanics.* John Wiley and Sons, NY 2011.
- Carey, G.F. and Oden, J.T., *Finite Elements: VI, Special Problems in Fluid Mechanics,* Prentice Hall Publishing Company, Englewood Cliffs, 1986.
- Oden, J.T. and Reddy, J.N., *Variational Methods in Theoretical Mechanics*, Springer-Verlag, New York, Heidelberg, Berlin, 1976.
- Oden, J.T. and Reddy, J.N., *An Introduction to the Mathematical Theory of Finite Elements*, John Wiley & Sons, NY, 1976.
- Oden, J.T., *Applied Functional Analysis: a First Course for Students of Mechanic and Engineering Students.* Englewood Cliffs, NJ, 1979.
- Oden, J.T., *Finite Elements of Nonlinear Continua*, McGraw-Hill Book Company, New York, 1972.
- Oden, J.T., *Mechanics of Elastic Structures*, McGraw-Hill Book Company, New York, 1967.

L. Gary Leal (Chair: 1994–1996)

Gary Leal was born in Bellingham, Washington on March 18, 1943. His degrees are in Chemical Engineering, B.S. University of Washington (1965), M.S. Stanford (1968), and Ph.D. Stanford (1969). Following his Ph.D., he spent a year as a Postdoctoral Fellow in Applied Mathematics at the University of Cambridge. From 1970 to 1989, he was in Chemical Engineering at Caltech where he rose through the ranks to Professor. He joined the University of California, Santa Barbara in 1989 as Chair of Chemical Engineering. He has served two terms as Chair of Chemical Engineering, and also has appointment in Materials and Mechanical Engineering. Leal is currently the Warren & Katharine Schlinger Professor of Chemical Engineering at UC Santa Barbara.

Leal is a member of the National Academy of Engineering (1987), Fellow of the American Academy of Arts and Sciences (2011), and has been honored with the APS Fluid Dynamics Prize (2002), the SOR Bingham Medal (2000), the AIChE William H. Walker Award (1993), and the AIChE Allan Colburn Award (1978). His research is concerned with the dynamics of complex fluids with emphasis on the coupling between flow and microstructure of polymeric liquids. He has been Editor of the *Journal Physics of Fluids, American Institute of Physics*, from 1998 to present.

Leal is the author of two text books: Laminar Flow and Convective Transport Processes (1992), and Advanced Transport Phenomena: Fluid Mechanics and Convective Transport Processes (2007).

Earl H. Dowell (Chair: 1996–1998)

Earl Dowell was born in Macomb, Illinois on November 16, 1937. He received a B.S. from the University of Illinois (1959), M.S. from MIT (1961), and Sc.D. from MIT (1964). His degrees are in aerospace engineering. He is the William Holland Hall Professor of Mechanical Engineering at Duke University. Dowell was the Dean of Engineering at Duke from 1983 to 1999, and in 2012 was appointed Chair of Mechanical Engineering and Materials Science at Duke.

Dowell is a member of the National Academy of Engineering (1993). His awards include: the AIAA/ASME/AHS/SAE Daniel Guggenheim Medal (2008), the ASME Den Hartog Medal, Lyapunov and Spirit of St. Louis Awards, the AAM Distinguished Service Award, and for the AIAA he is an Honorary Fellow, a Theodore Van Kármán Lecturer, and recipient of the Structures, Structural Dynamics, and Materials Award (1980) and the Walter J. and Angelina J. Crichlow Trust Prize.

Dowell's principal research activities are in the fields of aeroelasticity, acoustics, nonlinear dynamics, structural dynamics, and unsteady aerodynamics.

Ronald J. Adrian (Chair: 1998–2000)

Ron Adrian was born on June 16 in 1945. He received a B.M.E. (1967) and M.S. (1969) in Mechanical Engineering from the University of Minnesota, and a Ph.D. (1972) in Physics from Churchill College, Cambridge University. He was a member of Department of Theoretical and Applied Mechanics at the University of Illinois, Urbana-Champaign prior to joining the faculty at Arizona State University as the Ira A. Fulton Professor of Mechanical and Aerospace Engineering, Regents' Professor and head of the Laboratory for Energetic Flow and Turbulence.

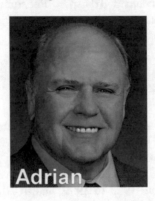

Adrian is a member of the United States National Academy of Engineering (1996). Adrian's awards include: the 2001 Nusselt-Reynolds Prize from the Assembly of World Conferences on Experimental Heat Transfer, Fluid Mechanics, and Thermodynamics, the 2002 AIAA Aerospace Measurement Technology Award, the 2005 APS Fluid Dynamics Prize, the 2007 AIAA Fluid Dynamics Award, the 2010 ASME Fluids Engineering Award, the 2009 Miegunyah Distinguished Fellowship, given by the Univ. Melbourne, and the 2010 Leonardo Da Vinci Award, given by the International Symposium for Flow Visualization.

He has served as an Associate editor of the Journal of Fluid Mechanics, a Co-editor of the Springer Series in Experimental Fluid Mechanics and Co-founder and editor of eFluids.com.

Adrian's research interests are the space–time structure of turbulent fluid motion and the development of techniques, both experimental and mathematical, to explore this structure. He co-edited Experiments in Fluids, and a ten volume series on Laser Techniques in Fluid Mechanics. He served as Chairman of the American Physical Society Division of Fluid Dynamics.

Hassan Aref (Chair: 2000–2002)

Hassan Aref was born on September 28, 1950 in Alexandria, Egypt. He was educated at the University of Copenhagen Niels Bohr Institute, graduating in 1975 with a C.S.D in Physics and Mathematics. Subsequently he received the

Ph.D. degree in Physics from Cornell University in 1980. Aref started his faculty career in the Division of Engineering at Brown University where he served from 1980 to 1985. He was in the faculty of University of California, San Diego, with appointments in the Department of Applied Mechanics and Engineering Science, and the Institute of Geophysics and Planetary Physics, 1985–1992. Additionally, he was Chief Scientist at the San Diego Supercomputer Center for 3 years, 1989–1992. Aref was the Head of the Department of Theoretical and Applied Mechanics at University of Illinois at Urbana-Champaign from 1992 to 2003, and was the Dean of Engineering (2003–2005) and then Reynolds Metals Professor in the Department of Engineering Science and Mechanics at Virginia Tech until his untimely death in 2011. Hassan also served as the Niels Bohr Visiting Professor at the Technical University of Denmark during the last several years of his life.

Aref was recognized for having developed the concept of chaotic advection in fluid mechanics. He was honored with the APS Otto Laporte Award (2000) and the SES G. I. Taylor Medal (2011). He was Associate Editor of Journal of Fluid Mechanics 1984–1994, and founding editor with David G. Crighton of *Cambridge Texts in Applied Mathematics*.

Aref served as chair of the APS Division of Fluid Dynamics. He was a member of the Executive Committee of the Congress Committee of the International Union of Theoretical and Applied Mechanics (IUTAM), a member of the National Academies Board on International Scientific Organizations, and a member of the Board of the Society of Engineering Science.

Aref was the president of the 20th International Congress of Theoretical and Applied Mechanics held in Chicago in 2000. This was the third time that the international congress had been held in the United States. Aref organized a consortium led by the University of Illinois, Urbana-Champaign, as the hosts for the Chicago Congress. He wrote and directed a mini play that interrupted his welcoming remarks during the opening ceremony in Chicago. As Aref spoke to the assembled international audience, Newton, Galileo and Archimedes (all dressed in the full garb of their time) came down separate aisles of the auditorium arguing loudly, across the audience, as to the importance of their contributions. This was an excellent example of the sense of humor, intellect, and imagination that Aref demonstrated in his everyday life.

Aref could also be profoundly serious at other times. Following is his statement of the Chair's report at the opening of the 2002 meeting of USNC/TAM:

Chair's report—Hassan Aref—June 23, 2002—Blacksburg, VA

The year and a bit since we last met has probably been one of the more dramatic periods while I have been coming to these meetings. Last year we were still buoyed, at least in part, by the strong economy. The main concerns of the country were with mechanical problems such as when a chad can be said to be disconnected from a ballot, and how to cleanly define the difference between a dimpled chad and a hanging chad, and other such profound issues.

The tragic events of 9/11 changed all that. The worsening economy was suddenly seen against the stark background of global terrorism. The U.S. was shaken by a violent attack on some of its institutions and landmarks, a first attack by foreigners on the U.S. mainland. 9/11 was likened to Pearl Harbor. There was grief. There was panic. There was an air of despair as the entire premise of an open society seemed to come under siege. The engineering profession, which many of us are part of or at least close to, was quick to understand what had happened. Suddenly, the ever present parameter called "safety" had been upgraded to "security against dedicated, planned, attack, with all the usual ideas and conventions about hostage taking and the value of human life suddenly pushed off to the sidelines."

Killing of civilians *en masse* and suicide attackers using civilian aircraft as weapons were probably things that some had thought about. But they were generally dismissed assuming that the moral price any attacker who used such methods would pay would to be too steep. That all changed on 9/11. Issues such as bioterrorism and dirty nuclear weapons aimed at civilian populations were openly discussed in the media. As life returned to normalcy, we endured months of heart-wrenching pictures from downtown New York, now known as Ground Zero. Travel changed dramatically—and those of us with Arabic names got to sample many of the embellishments. On top of that the country went to war and economic optimism quickly plummeted. In the ensuing months the bad news continued, including things like the Enron scandal and then the dissolution of Arthur Andersen, once thought to be about as golden as one could get. The great budget surpluses of a year or two ago evaporated. On the mechanics front we have witnessed extensive flooding in several parts of the country, and now raging wildfires in the Southwest.

The bitterness and disillusionment that set in has, I think, affected every one of us. At my home institution the Federal budget woes have been mirrored in State budget woes, and since we are a public institution, in University budget woes. I am sure my colleagues in the public universities elsewhere know exactly what I am talking about, because we are not alone. Our colleagues in the private schools have been somewhat sheltered from this fall-out, but surely have not failed to notice the mood change in the country at large. I say all this not because you don't know it, probably in some cases you know it even better than I. I sincerely hope that none of you have had personal losses from the tragedies of 9/11, and if you have, I offer my heartfelt condolences. But even if you haven't, I suspect you all feel a sense of loss, grief, maybe anger and frustration, certainly unease. I can testify for my own part that much of my time over the past 9 months has been preoccupied with dealing with the repercussions of things happening well outside my University. The world of science to which we belong is very international, maybe not with all that much representation from the part of the world most directly involved in the aftermath of 9/11, but we certainly see plenty of students from the Middle East, and we have lots of dialog with the scientific community of Israel. We have connections with scientists from Pakistan to some extent, and certainly with the scientific community in India at all levels. We have students and colleagues from Russia and the former Soviet Union, many of whom have quite different perspectives on much of what is happening in the world today. We have individuals with all kinds of religious beliefs and ideologies in our universities. Some of these beliefs are technically at war with one another elsewhere in the world.

Hassan Aref passed away on Sept. 9, 2011, at age 60, sitting in his chair, in De Land, Illinois, USA.

Wolfgang G. Knauss (Chair: 2002–2004)

Wolfgang Knauss was born on December 12, 1933 in Mandel bei Bad Kreuznach, Germany. Following graduation in 1954 from the Helmholtz—Real Gymnasium in Heidelberg, he immigrated to the United States to study, first for 1 year at Pasadena City College, and then at Caltech for the remainder of his studies.

He earned his B.S. (1958), M.S. (1959), and Ph.D. (1963), all from Caltech. Following the Ph.D., he was a Research Fellow in Aeronautics at Caltech for 2 years, and then became the Assistant Professor of Aeronautics and Applied Mechanics. He rose through the ranks eventually becoming the Theodore von Kármán Professor of Aeronautics and Applied Mechanics. He remained at Caltech throughout his career and is now professor emeritus. Knauss became an American citizen in 1967.

Knauss has made fundamental contributions, with emphasis on experimental work, in the broad area of viscoelastic material behavior. He has received numerous awards including: National Academy of Engineering (1998); ASME Koiter Medal (2001); ASME Timoshenko Medal (2010); National Academy of Sciences Lecturer to the U.S.S.R. (1977); Senior U.S. Scientist Award by the Alexander von Humboldt Foundation (1986/87); SEM Murray Medal (1995), Lazan Award (2000) and SEM Honorary Member (2002); Foreign Member of the Russian Academy of Natural Sciences (1997); Corresponding Member of the International (Russian) Academy of Engineering, and recipient of its Kapitsa Medal (1996).

Knauss' academic genealogy traces back to Otto Mohr of Mohr's Circle fame, in chronological order via Max L. Williams, Ernest E. Sechler, Theodore von Kármán, Ludwig Prandtl, August Föppel and finally, Christian Otto Mohr.

Ted Belytschko (Chair: 2004–2006)

Ted Belytschko was born on January 13, 1943 in Lviv, Ukraine. He immigrated to the United States in 1951 settling in Chicago, and became a U.S. citizen in 1961. He received a B.S. in Engineering Sciences (1965) and Ph.D. in Mechanics (1968) from the Illinois Institute of Technology. Following his doctorate under Phil Hodge, Ted accepted a position as Assistant Professor of Structural Mechanics at the University of Illinois at Chicago. He spent 9 years there, rising through the ranks to Full Professor. He then moved to Northwestern University in 1977 as Professor of Computational Mechanics. At Northwestern, Belytschko served as Chairman of Mechanical Engineering for 5 years, became Walter P. Murphy Professor of Computational Mechanics, and eventually McCormick Distinguished Professor. He also held appointment in the Civil and Environmental Engineering Department.

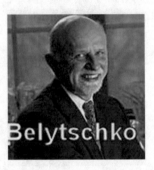

Belytschko's research was concerned with computational methods for engineering problems. He developed methods for incorporating arbitrary weak and strong discontinuities in finite element methods. He pioneered the concept of mesh-free methods, and developed explicit finite element methods widely used in crashworthiness analysis and virtual prototyping. Much of his work has been devoted to modeling the behavior of solids, with particular emphasis on failure and fracture over a range of scales. He made major contributions to the field and has been recognized for his efforts with numerous honors and awards. He is a past chair of the Applied Mechanics Division of ASME (1991), and is past Editor-in-Chief of the International Journal for Numerical Methods in Engineering.

Belytschko's awards include: Member of the National Academy of Science (2011); Member of the National Academy of Engineering (1992); USACM John von Neumann Medal (2001); ASME Timoshenko Medal (2001); Fellow of the American Academy of Arts and Sciences (2002); IACM Gauss-Newton Medal (2002); ASCE Theodore von Kármán Medal (1999) and Huber Research Prize (1977), SES William Prager Medal (2011); Melvin Baron Medal, from the Shock and Vibration Information Analysis Center (1999); USACM Structural Computational Mechanics Award (1997); ICES Medal (1997); JSME Computational

Mechanics Award (1993), ASCE Structural Dynamics and Materials Award (1990); Thomas Jaeger Prize, from the International Assn. for Structural Mechanics in Reactor Technology (1983), and ASME Honorary Member (2013).

ASME renamed the ASME Applied Mechanics Award as the Ted Belytschko Applied Mechanics Division Award in 2007. The USACM Structures award was renamed the Belytschko Medal for Computational Solid and Structural Mechanics in 2012. The Ted Belytschko Lecture at Northwestern University was established in Ted's honor in 2013.

Belytschko received Honorary Doctorates from the University of Liege (1997), Institute National des Sciences Appliquées de Lyon (2006), and École Normale Supérieure de Chacan, Paris (2004).

Ted Belytschko died on September 15, 2014.

Belytschko co-authored five books:

- T. Belytschko and T. J. R. Hughes, *Computational Methods for Transient Analysis*, North-Holland, Amsterdam (1983).
- W. K. Liu, T. Belytschko and K. C. Park, *Innovative Methods for Nonlinear Problems*, Pineridge Press, Swansea, UK (1984).
- W. K. Liu and T. Belytschko, *Computational Mechanics of Probabilistic and Reliability Analysis*, Elmepress International, Lausanne, Switzerland (1989).
- T. Belytschko, W. K. Liu, and B. Moran, *Nonlinear Finite Elements for Continua and Structures*, John Wiley & Sons, Ltd., Chichester, England (2000).
- J. Fish, T. Belytschko, *A First Course in Finite Elements*, John Wiley & Sons, Ltd., Chichester, England (2007).

Nadine Aubry (Chair: 2006–2008)

Nadine Aubry was born on March 14, 1960 in Nantes, France. She received a "diplome d'Ingenieur" (1984) from the National Polytechnic Institute of Grenoble, France, and a master's degree (1984) from the Scientific and Medical University of Grenoble. She earned her Ph.D. in 1987 from the Sibley School of Mechanical and Aerospace Engineering at Cornell University. After a year as postdoctoral fellow, Nadine accepted a faculty position at the Levich Institute and in Mechanical Engineering at City College of the University of New York. Thereafter she became Professor and Jacobus Chair at the New Jersey Institute of Technology where she rose to Mechanical Engineering Department Head in 2000 and Distinguished Professor in 2002. It is believed that she was the first female head of a mechanical engineering department in the U.S. In 2006, she moved to Carnegie-Mellon University, as the Head of Mechanical Engineering and became R. J. Lane Distinguished Professor in 2009 and University Professor in 2012. Nadine then moved to Northeastern University in 2012 as Dean of the College of Engineering.

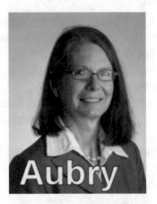

Aubry's research is in the area of fluid dynamics. She pioneered the modeling of open flow turbulence and other complex flows and systems using advanced decomposition techniques and dynamical systems theory. She also worked on flow control and made noteworthy contributions to the field of microfluidics, pioneering judicious micromixers, and particle manipulators.

Nadine is a member of the National Academy of Engineering (2011). Additional awards include: National Science Foundation Presidential Young Investigator Award, and the Ralph R. Teetor Educational Award from the Society of Automotive Engineers (SAE). She was elected Chair of the APS Division of Fluid Dynamics for the 2013–2014 term. In 2012 she was elected as the first female member of the IUTAM Bureau.

Aubry became a citizen of the United States in 2008.

Thomas J. R. Hughes (Chair: 2008–2010)

Tom Hughes was born in Brooklyn, New York on August 3, 1943. He received a B.E. (1965) and M.E. (1967) in Mechanical Engineering from Pratt Institute. He then earned M.S. (1974) in Mathematics and Ph.D. (1974) in Engineering Science from the University of California at Berkeley.

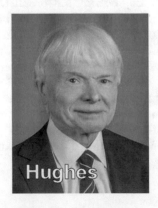

Hughes began his career as a mechanical design engineer at Grumman Aerospace (1965), subsequently joining General Dynamics as a research and development engineer. After receiving his Ph.D., he joined the Berkeley faculty, eventually moving to California Institute of Technology (1976) and then Stanford University (1980), before joining the University of Texas at Austin in 2002. At Stanford, he served as Chairman of the Division of Applied Mechanics, Chairman of the Department of Mechanical Engineering, and Chairman of the Division of Mechanics and Computation, and occupied the Mary and Gordon Crary Chair of Engineering.

At Texas, he is professor of Aerospace Engineering and Engineering Mechanics, holder of the Computational and Applied Mathematics Chair III, and leader of the ICES Computational Mechanics Group at the University of Texas (Austin).

Hughes' primary interest is computational mechanics with emphasis on isogeometric analysis (Cottrell et al. 2009), finite element analysis, variational multiscale methods for complex fluid flows and turbulence, and patient-specific biomedical modeling and simulation.

Hughes is a member of the U.S. National Academy of Sciences (2009), the U.S. National Academy of Engineering (1995), Fellow of the American Academy of Arts and Sciences (2007), the Royal Society of London, the Austrian Academy of Sciences (Section for Mathematics and the Physical Sciences), the Istituto Lombardo Accademia di Scienze e Lettere (Mathematics Section), and the Academy of Medicine, Engineering, and Science of Texas.

He is the co-editor of the international journal Computer Methods in Applied Mechanics and Engineering, a founder and past President of USACM and IACM, and past Chairman of the Applied Mechanics Division of ASME.

Hughes has been awarded the Timoshenko (2007), Worcester Reed Warner (1998), and Melville Medals from ASME (1979); the Walter L. Huber Civil Engineering Research Prize (1978) and the von Kármán Medal (2009) from ASCE; the von Neumann Medal from USACM (1997); the Gauss-Newton Medal from IACM (1998); the Computational Mechanics Award from JSME (1993); the Grand Prize from the Japan Society of Computational Engineering and Science (JSCES); and the Humboldt Research Award for Senior Scientists from the Alexander von Humboldt Foundation.

Hughes has received honorary doctorates from the Universite Catholique de Louvain, the University of Pavia, the University of Padua, the Norwegian University of Science and Technology (Trondheim), Northwestern University (Evanston), and the University of A Coruña. He held the Cattedra Galileiana (Galileo Galilei Chair), Scuola Normale Superiore, Pisa, in 1999, and the Eshbach Professorship, Northwestern University, in 2000.

The Special Achievement Award for Young Investigators in Applied Mechanics is an award given annually by the Applied Mechanics Division of ASME. In 2008 this award was renamed the *Thomas J.R. Hughes Young Investigator Award*. In 2012, the Computational Fluid Mechanics Award of the United States Association for Computational Mechanics was renamed the *Thomas J.R. Hughes Medal*.

Lance Collins (Chair: 2010–2012)

Lance Collins was born on September 9, 1959 in Westbury (Long Island), NY. He earned three degrees in Chemical Engineering, B.S.E Princeton (1981), and M.S. (1983) and Ph.D. (1987), University of Pennsylvania. Collins is now the Joseph Silbert Dean of Engineering at Cornell University. Prior to that, he served as the S. C. Thomas Sze Director of the Sibley School of Mechanical & Aerospace Engineering from 2005 to 2010, and he was the Director of Graduate Studies for Aerospace Engineering 2003–2005. Collins joined Cornell in 2002, following 11 years as Assistant Professor, Associate Professor, and Professor of Chemical Engineering at the Pennsylvania State University. From 1999 until his departure, he also held a joint appointment in the Mechanical & Nuclear Engineering Department at Penn State.

In 1998 Collins was a visiting scientist at the Laboratoire de Combustion et Systémes Réactifs (a National Center for Scientific Research laboratory in Orleans, France) and at Los Alamos National Laboratory. In 2011, Dean Collins was part of the Cornell leadership team that successfully bid to collaborate with New York City to build a new campus on Roosevelt Island focused on innovation and commercialization in the tech sector. His research combines simulation and theory to investigate a broad range of turbulent flow processes. A unifying theme in his work is the importance of small-scale (microturbulence) transport. A second focus is on developing a new class of turbulence models that are capable of describing microturbulence processes.

Honors and awards that Collins has received include: Dow Chemical Young Minority Investigator Award (1990), Outstanding Paper Award, AIChE (1997), Cornell's Stephen Miles Outstanding Teaching Award (2004), Member-at-Large (Division of Engineering & Physical Sciences (Committee of the National Academies)) 2011, Executive Committee (APS Division of Fluid Dynamics) 2008–2011.

Chapter 11
Secretaries of the USNC/TAM

For any organization, the position of secretary is critically important for maintaining contact with the members as well as securing and maintaining records for future generations. The secretaries who have served the USNC/TAM and their periods of service are:

- 1948–1958   C. E. Davies (ex-officio from ASME)
- 1958–1970   O. B. Schier II (ex-officio from ASME)
- 1970–1982   F. N. Frenkiel (1st member of the committee as secretary)
- 1982–2000   P. G. Hodge, Jr. (Nov. 1, 1982–Oct. 31, 2000)
- 2000–2012   C. T. Herakovich (Nov. 1, 2000–Oct. 31, 2012)

Background on the secretaries is presented in the following.

Davies and Schier (Ex-officio Secretaries: 1948–1970)

The support to the committee by ASME during the years 1948–1970 is evidenced by the services provided by the then Secretaries of ASME, C. E. Davies, 1948–1958, and O. B. Schier, 1958–1970. Both men were ex-officio members of the committee, serving as secretary during the time that some meetings were held at ASME Headquarters in New York. During this period, the Secretary of ASME was the Chief Administrator Officer of the society. Later the position was given the title Executive Director of ASME. It is inconceivable today to think that the Executive Director of ASME would serve as secretary of the USNC/TAM. Nevertheless, this does demonstrate the very close relationship between ASME and the USNC/TAM during the early years of the committee.

There are no records of minutes in the USNC/TAM files and ASME has not been able to locate minutes of these meetings for the period 1948–1970. All efforts to locate minutes during the time when the Secretary of ASME acted in an ex-officio

© Springer International Publishing Switzerland 2016
C.T. Herakovich, *Mechanics IUTAM USNC/TAM*,
DOI 10.1007/978-3-319-32312-1_11

capacity as the Secretary of the USNC/TAM have been unsuccessful. This is the case even though, Dave Soukup at ASME headquarters has made serious efforts to locate minutes in ASME storage facilities, and email requests have been sent to all living former USNC/TAM members who might have copies of old minutes.

François N. Frenkiel (Secretary: 1970–1982)

As stated in a previous chapter, little is known about the committee during the early period (1948–1976). What is known can be found in the report by Paul Naghdi on the *History of the ASME Applied Mechanics Division*. The report was written for the 50th Anniversary of the Division; it was published in the Journal of Applied Mechanics in 1979. Information on the USNC/TAM was supplied to Naghdi by Francois Frenkiel in a document dated August 31, 1978. Frenkiel provided hand-written notes to Naghdi who then had them typed and signed by Frenkiel. Frenkiel was secretary of USNC/TAM when he wrote these notes. In addition to general statements concerning the formation of the committee, Frenkiel provided a complete listing of the committee chairs from 1949 to 1980, and a listing of the U.S. National Congresses of Applied Mechanics from 1951 to 1978. While Frenkiel started his list of chairs at 1949, it is clear that Dryden was the chair in 1948.

Frenkiel took over the secretary job in 1970 and served as secretary until 1982. He was the first secretary of the committee who was a voting member of the committee and not an ex-officio member from ASME. Minutes are available for 5 of the 12 years of Frenkiel's tenure as Secretary; the minutes for three of these years were taken by other individuals as Frenkiel was not present at those meetings. The available minutes during Frenkiel's tenure are: May 1976 and January 1979, taken by Frenkiel; October, 1979, taken by George Handelman; March and August, 1980 taken by Frenkiel; November 1981, taken by Dick Christensen; and June 1982 taken by Dick Skalak.

While Frenkiel did miss some of these meetings, when I asked Dick Christensen why he took the minutes in 1982, he told me Frenkiel "was extremely highly regarded and equally highly conscientious. I do not remember why he wasn't there, or that I took notes. My unsubstantiated recollection would be illness. I remember him very favorably but that was so long ago."

When I asked Hodge about his initiation to the secretary position, he responded: "When I started the job, I had nothing to go on. The previous secretary had accepted one more term than he should have. He was in ill health and was of no help to me whatever—I essentially had to redefine the position." Hodge's statement is consistent with Christensen's to the effect that Frenkiel was not in good health in the latter years of his tenure as secretary.

As Frenkiel was also chair of the USNC/TAM from 1964 to 1966, details of his professional career are described under the chapter on USNC/TAM Chairs. When Frenkiel completed his service to the committee, he was honored with a resolution (Appendix F) recognizing his contributions.

Philip G. Hodge, Jr. (Secretary: 1982–2000)

Phil Hodge was elected Secretary of the committee at a June 21–22, 1982 meeting. His first 2-year term began on Nov. 1, 1982. Hodge had not been a member of the committee prior to his election as Secretary. Hodge's tenure as Secretary provided the committee with the administrative continuity and stability that resulted in the most complete set of records being available for posterity.

With the exception of the missing minutes from 1977 to 1978 (prior to Hodge's tenure), Hodge passed on a complete set of committee documents from 1976 through the year 2000 when he retired as secretary. Minutes and other committee documents such as history, constitution, by-laws, and manual of operations from the Hodge years are available as paper copy.

Phil Hodge was the longest serving secretary of the USNC/TAM. His 18 years of service exceeded the next longest serving secretary by 6 years. He was meticulous in his work and put the committee on a firm administrative foundation. In addition to maintaining excellent, detailed minutes and correspondence with committee members, he was the lead writer of a Manual of Operations detailing the duties of each officer and each subcommittee; he also compiled a detailed history of the committee membership, as best he could from the records available at the time. In an email to me years ago, Ben Freund said that Hodge was responsible for putting the USNC/TAM on a sound organizational footing.

In addition to his secretarial duties, Hodge was a U.S. Representative to the IUTAM General Assembly from 1982 to 2000 and an IUTAM Member-at-Large from 2000 to 2008. When he retired from the secretarial duties, the committee expressed its appreciation and admiration to Hodge with a Resolution (Appendix F) and a plaque recognizing "his dedicated and outstanding service to the committee."

Philip Hodge was born in New Haven, Connecticut, on November 9, 1920. He earned a B.A. in Mathematics from Antioch College (Yellow Springs, Ohio) in 1943, and a Ph.D. in Applied Mathematics from Brown University in 1949. His Ph.D. advisor was William Prager; his thesis was concerned with torsion of plastic bars. He held teaching positions at UCLA (1949–1953), Polytechnic Institute of Brooklyn (1953–1957), Illinois Institute of Technology (1957–1971) and University of Minnesota (1971–1991) and visiting professor emeritus at Stanford in retirement. Hodge served in the U.S. Merchant Marines during the Second World War.

Hodge served ASME in a variety of capacities. He was the Technical Editor of the ASME Journal of Applied Mechanics from 1971 to 1976. He was a member of the Executive Committee of the Applied Mechanics Division (1963–1968) and its chair in 1968. He served in the ASME Basic Engineering Department Policy Board (1969–1974), National Nominating Committee (1971–1973, Chair in 1972–1973); Honors Committee (1978–1986); Constitution and By-Laws Committee (1986–1992), Committee on Rules (1992–1996); Committee on Planning and Organization (1996–1998).

ASME honored Hodge with the Worcester Reed Warner Medal (1975), Honorary Membership (1977), ASME Medal (1987), and the ASME Drucker Medal (2000). He was awarded the ASCE von Kármán Medal (1985) and the Euler Medal of the USSR Academy of Sciences (1983). He received the American Academy of Mechanics Award (1984), and was elected to the National Academy of Engineering in 1977.

Hodge's research resulted in significant advancements in plasticity theory including developments in the method of characteristics, limit-analysis, piecewise linear isotropic plasticity, and nonlinear programming applications. Hodge was one of the first to use finite element methods in his work. He authored five books, one co-authored with William Prager, and another co-authored with J. N. Goodier.

In addition to his academic endeavors, Hodge was a mountain climber (Mt. Ritter in California), a marathon runner (Boston) and a fan of live theater. In retirement, he wrote opera reviews. He died in Sunnyvale, California on November 11, 2014 at age 94.

Hodge Books:

- Hodge, Jr., P. G., (1963), *Limit Analysis of Rotational Symmetric Plates and Shells*, Prentice-Hall, Inc. Englewood Cliffs, NJ.
- Hodge, Jr., P. G., (1959), *Plastic Analysis of Structures*, McGraw-Hill book Co., New York, NY.
- Hodge, Jr., P G., (1970), *Continuum Mechanics*, McGraw-Hill book Co., New York, NY.
- Prager, W. and Hodge, Jr., P G., (1951), *Theory of Perfectly Plastic Solids*, John Wiley & Sons, London.
- Goodier, J. N. & Hodge, Jr., P G., (1958), *Elasticity and Plasticity*, John Wiley & Sons, New York, NY.

Carl T. Herakovich (Secretary: 2000–2012)

Carl Herakovich succeeded Phil Hodge as secretary of the USNC/TAM on November 1, 2000. At that time Hodge turned over all the records he had in his possession, including copies of minutes, constitution, by-laws and manual of operations. Continuing in the style developed by Hodge, Herakovich converted the committee's secretarial efforts into an electronic form. Essentially all communications after 2000 were made via email. As a result, all committee documents for the years 2000–2012 now are available in electronic form. These electronic files as well as all paper files passed on by Hodge were turned over to the secretary Linda Franzoni who took over in 2012.

The electronic records also were transmitted to the National Academy of Sciences as they were developed. In addition, Herakovich turned over to Franzoni many volumes of Proceedings of the U.S. National Congresses on Applied Mechanics that he had obtained over the years. When he retired from the secretarial duties, the committee expressed its appreciation and gratitude to Herakovich with a plaque (Appendix F) recognizing "his dedicated and outstanding service to the committee."

Carl Herakovich was born in East Chicago, Indiana, on August 6, 1937. He lived in the adjacent town of Whiting, Indiana (a Chicago suburb) and attended Whiting High School graduating in 1955. He then received a B.S., in Civil Engineering (1959) from Rose Polytechnic Institute (now Rose-Hulman Institute of Technology). Following a year in which he completed active duty military service in the Army Corps of Engineers and civilian work as an engineer, he entered the University of Kansas where he received a M.S., in Mechanics (1962). Near the completion of his M.S., he was invited to return to Rose-Hulman as the Head Football Coach and Instructor in Civil Engineering. Soon thereafter, he was appointed as Athletic Director. He served in these capacities for 2 years. In September 1964, he entered the Illinois Institute of Technology (IIT) officially receiving the Ph.D., in Mechanics in 1968. Phil Hodge was his Ph.D. advisor.

During and following his graduate education, Herakovich taught mechanics courses at the University of Kansas, Rose-Hulman Institute of Technology, Illinois Institute of Technology, Virginia Polytechnic Institute and State University (Virginia Tech), and the University of Virginia. While at Virginia Tech, Herakovich conceived the idea of and directed the NASA-Virginia Tech Composites Program. He also led the movement to change the name of the Engineering Mechanics Department to the Department of Engineering Science and Mechanics (ESM) and the movement to require personal computers for all entering freshman engineering students in 1984. At Virginia, he was the Henry L. Kinnier Professor and Director of the Applied Mechanics Program in the School of Engineering and Applied Science. He retired from Virginia in 1968 as the Henry L. Kinnier Professor Emeritus.

Herakovich was Chair of the ASME Applied Mechanics Division (1996–1997) and ASME Vice-President of Basic Engineering (2001–2004). He served on the SES Board of Directors, (1983–1992), and was President in 1992. Herakovich joined the USNC/TAM in 1996 as the ASME representative. He was a U.S. representative to the IUTAM General Assembly (1998–2016). He also served as Chair of the IUTAM Working party on Mechanics of Materials (2006–2007) and as a member of the IUTAM Congress Committee (2006–2014).

Herakovich was honored with the Ted Belytschko Award of the ASME Applied Mechanics Division (2005), SES Fellow (2006), IIT Mechanical, Materials & Aerospace Engineering Department, Alumni Recognition Award (2010), and Virginia Tech ESM, Frank Maher Outstanding Educator Award (1986).

Herakovich's primary research interest is mechanics of composite materials with emphasis on theoretical modeling and experimental correlation. He is the author of the text *Mechanics of Fibrous Composites* (Herakovich, 1998) and an eBook on *Elastic Solids* published in 2013. This appears to be the first eBook ever published with a full set of mathematical equations.

In addition to his career in mechanics, Herakovich has had an unusual athletic career for an engineering professor. He was a quarterback and defensive halfback on the 1954 Indiana High School Football Championship team, leading collegiate scorer in the United States in 1958 when he scored 168 points in eight games, also was a football coach and athletic director at Rose-Hulman in 1962 and 1963, and Atlantic Coast Conference football official for 20 years (1972–1991). Herakovich was elected to the Indiana Football Hall of Fame (1985), the Rose-Hulman Athletic Hall of Fame (1993), and the School City of Whiting Buckley Wall of Fame (2005).

Chapter 12
USNC/TAM in IUTAM Leadership Positions

Complete listings of USNC/TAM members who served IUTAM as Members-at-Large, Members of the Bureau, Members of the Congress Committee, Members of the Executive Committee of the Congress Committee, and Members of the Fluids and Solids Symposium Panels are provided in Appendix G. Following is a summary of the leadership positions held by USNC/TAM members. A brief bio is included for those individuals whose background has not been described previously.

Hugh L. Dryden (IUTAM 1948–1960)

Hugh Dryden served as the first IUTAM Treasurer (1948–1952), President (1952–1956), and Vice-President (1956–1960). See committee chairs for additional information.

Nicholas J. Hoff (IUTAM 1948–1998)

Nick Hoff served as a member of the IUTAM Bureau from 1960 to 1968. See committee chairs for additional information.

© Springer International Publishing Switzerland 2016
C.T. Herakovich, *Mechanics IUTAM USNC/TAM*,
DOI 10.1007/978-3-319-32312-1_12

Bruno A. Boley (IUTAM 1964–2014)

Bruno Boley was a member of the IUTAM Bureau as Treasurer from (1992 to 1996). He was a member of the IUTAM Congress Committee (CC) from 1968 to 1996, member of the Executive Committee of the CC (1972–1982), and Secretary of the CC (1976–1983). See committee chairs for additional information.

Daniel Drucker (IUTAM 1960–1988)

Dan Drucker served on the IUTAM Bureau from 1960 to 1988. He was Bureau Member (1960–1972), Treasurer (1972–1980), President (1980–1984), and Vice-President (1984–1988). See committee chairs for additional information.

L. Ben Freund (IUTAM 1986–2016)

Ben Freund joined USNC/TAM in 1985 as a Member-at-Large. He was elected as U.S. Representative to the IUTAM General Assembly in 1986. Ben served IUTAM in several capacities during his tenure. He is on schedule to complete 31 years of continuous service to the community in 2016. He was a U.S. Representative to the IUTAM General Assembly (1986–1996), IUTAM Treasurer (1996–2004), President (2004–2008), and Vice-President (2008–1912). Freund began a 4-year term as a Member-at-Large to the IUTAM General Assembly in 2012.

Ben Freund was born on November 23, 1942, in Johnsburg, Illinois. He earned B.S. and M.S. Degrees in Engineering Mechanics at the University of Illinois in 1964 and 1965, respectively. He earned his Ph.D. in Theoretical and Applied Mechanics from Northwestern University in 1967. Ben's Ph.D. advisor at Northwestern was Jan Achenbach. Following his graduate studies, he joined Brown University, first as a postdoctoral fellow and then as a member of the faculty. At Brown, he rose through the ranks, eventually becoming the H. L. Goddard University Professor. Freund took emeritus status at Brown in 2010. He currently is an Adjunct Professor of Materials Science and Engineering at the University of Illinois.

Freund served as Chair of the Executive Committee of the Applied Mechanics Division of ASME (1994), Editor of the ASME Journal of Applied Mechanics (1983–1988), Editor of the Journal of Mechanics and Physics of Solids (1992–2005), and Editor of Cambridge Monographs on Mathematics and Mechanics (1992–2004).

His research is concerned with stress waves in solids, fracture mechanics, seismology, dislocation theory, mechanics of thin films, and bioadhesion. He is the author of two monographs, one on *Dynamic Fracture Mechanics* (Freund, 1990) and, a second with S. Suresh, on *Thin Film Materials* (Freund and Suresh, 2003).

Freund has been honored with the George R Irwin Medal of ASTM (1987), William Prager Medal of SES (2000), Stephen P Timoshenko Medal of ASME (2003), and Arpad Nadai Medal of ASME (2009). He is a member of the National Academy of Engineering (1994) and the National Academy of Sciences (1996), and Fellow of American Academy of Arts and Sciences (1993).

Nadine Aubry (IUTAM 2007–2016)

Nadine Aubry was a U.S. Representative to the IUTAM General Assembly (2007–2013) and elected to a 4-year term on the IUTAM Bureau in 2012. Nadine is the first female ever elected to the IUTAM Bureau. See committee chairs for additional information.

Hassan Aref (IUTAM 1992–2011)

Hassan Aref was a member of the IUTAM Congress Committee (1992–2011) and Secretary of the Congress Committee from 2008 until his death in 2011. He also was General Chair of the very successful 20th International Congress in Chicago. See committee chairs for additional information.

Robert McMeeking (IUTAM 2008–2016)

Bob McMeeking was elected a Member-at-Large of the USNC/TAM in 2007 and remains on the committee as a result of being elected Secretary of the IUTAM Congress Committee for a 4-year term that began in 2012. The Secretary of the Congress Committee is responsible for the overall organization and coordination of the international congress, working in cooperation with the local organizing committee.

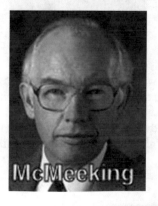

McMeeking was born on May 22, 1950 in Glasgow, Scotland. He received a B. Sc. In Mechanical Engineering from the University of Glasgow in 1972, His M.S. (1974) and Ph.D. (1977) are in Solid Mechanics from Brown University. He has held faculty positions at Stanford (1976–1978), Illinois (1978–1985) and UC Santa Barbara since 1985.

McMeeking's research is concerned with solid mechanics, materials, and structures to include: mechanics of materials, fracture, composite materials, materials processing, thermal barrier coatings, and biomechanics.

His honors include: SES Prager Medal (2014), National Academy of Engineering (2005), Alexander von Humboldt Research Award (2004 and 2013), Editor-in-Chief, ASME Journal of Applied Mechanics (2002–2012), ASME Timoshenko Medal (2014).

Jan D. Achenbach (IUTAM 1980–2016)

Jan Achenbach was a Member-at-Large of the USNC/TAM from 1973 to 1981. He then served on the IUTAM International Papers Selection Committee for the 15th International Congress (1980) and on the IUTAM Solids Symposium Panel (1982–2008). He was Chair of this Panel (2004–2008). Achenbach also was a U.S. Representative to the IUTAM General Assembly for the 1996–1998 term. In recognition of all his contributions to IUTAM, the USNC/TAM, and his standing in the mechanics community, Achenbach was elected as a Member-at-Large to the IUTAM General Assembly in 2008 and has been re-elected several times and will serve at least through 2016.

Jan Drewes Achenbach was born in Leeuwarden, the Netherlands, on August 20, 1935. He studied aeronautics at Delft University of Technology completing a M.Sc. degree in 1959. He then went to Stanford University where he received his Ph.D. in 1962. C. C. Chao was his advisor at Stanford. After working for a year as a postdoctoral fellow at Columbia University, he accepted a position as assistant professor of civil engineering at Northwestern University and later accepted a joint appointment in Mechanical Engineering. He rose through the ranks at Northwestern and retired as the Walter P. Murphy Professor and Distinguished McCormick School Professor, emeritus. Achenbach's research is both analytical and experimental. He has been concerned with methods for flaw detection, dynamic fracture, structural acoustics, and mechanical behavior of composite materials.

Achenbach is the founding Editor-in-Chief of *Wave Motion*, and was Editor-in-Chief from the journal's establishment in 1979 until 2012. His work has been recognized with major awards including: ASEE Curtis W. McGraw Research Award (1975), McDonnell-Douglas Aerospace Model of Excellence Award (1996). ASME Timoshenko Medal (1992), ASME Honorary Member (2002), ASME Medal (2012), SES William Prager Medal (2001), National Medal of Technology (2003), National Medal of Science (2005), ASCE Raymond Mindlin Medal (2009), ASCE Theodore von Kármán Medal (2010), and AAM Outstanding Service Award (1997).

Additional honors include: National Academy of Engineering (1982), National Academy of Science (1992), and Corresponding Member, Royal Dutch Academy of Sciences (1999). He is a Fellow, American Academy of Arts and Sciences (1994), American Association for the Advancement of Science (1994), and Japan Society for the Promotion of Science (1982). He was elected a fellow of the World Class Universities Program of the National Research Foundation of Korea (2009), and received an Honorary Doctorate from China's Zhejiang University (2011).

Achenbach has authored or co-authored four books:

- Achenbach, J. D. (1973). *Wave propagation in elastic solids*. North-Holland Series in Applied Mathematics and Mechanics. North-Holland. ISBN 0-7204-0325-1.
- Achenbach, J. D. (1975). *A theory of elasticity with microstructure for directionally reinforced composites*. Springer. ISBN 978-3-211-81234-1.
- Achenbach, J. D. Gautesen, A. K.; McMaken, H. (1982). *Ray methods for waves in elastic solids*. Pitman Advanced Pub. Program. ISBN 0-273-08453-4.
- Achenbach, J. D. (2003). *Reciprocity in elastodynamics*. Cambridge University Press. ISBN 978-0-521-81734-9.

L. Gary Leal (IUTAM 1992–2016)

Gary Leal was a U.S. Representative to the IUTAM General Assembly (1992–1998) and (2000–2009). He began service as a member of the IUTAM Fluids Symposium Panel in 2000. He was appointed Chair of the panel for two terms (2008–2016). Gary was also a member of the IUTAM Congress Committee from 2000 to 2008. See Committee Chairs for additional information.

Chapter 13
U.S. Members of the IUTAM General Assembly

The U.S. is authorized five Representatives (based upon the level of dues the U.S. pays), a variable number Member-at-Large positions (based upon the vote of the General Assembly) and has always had one additional member of the Bureau in the General Assembly. The Members-at-Large initially were referred to as Elected or Personal Members. In 1950, a new policy restricted the number of elected members, but introduced a new class of members who were Representatives from the Adhering countries. The listing of the U.S. members of the IUTAM General Assembly in Appendix G is based on the IUTAM Annual Reports. In the listing, Bureau Members are denoted as follows: T - Treasurer, P - President, V - Vice-President, B -Bureau Member, HP - Honorary President.

© Springer International Publishing Switzerland 2016
C.T. Herakovich, *Mechanics IUTAM USNC/TAM*,
DOI 10.1007/978-3-319-32312-1_13

Chapter 14
USNC/TAM Diversity

A variety of diversity issues come into play for a U.S. National Committee. Over the years, the USNC/TAM has paid considerable attention to several of these including mechanics specialty, age, and organizational representation. Other types of diversity such as gender, racial, and ethnic have received attention only in more recent years. The committee has been unusually successful in addressing the diversity question. Following are statements and data that reflect the Committee diversity through 2012.

Academic Genealogy

Appendix I presents a listing of the 191 known members of the USNC/TAM through October 31, 2012. The list shows the 56 universities where members earned their doctorates, and the names of their 162 university advisors. The members and their advisors represent an all-star listing of mechanicians who have contributed to the field. The universities where members earned their doctorates are from 14 different countries. Clearly, the committee members are from very diverse academic backgrounds. The highest numbers of doctorates from a university are: Caltech—23, MIT—16, Illinois—11, Brown—10, and Stanford—10.

© Springer International Publishing Switzerland 2016
C.T. Herakovich, *Mechanics IUTAM USNC/TAM*,
DOI 10.1007/978-3-319-32312-1_14

Gender Diversity

There have been 178 male members and 13 female members (7 %) of the USNC/
TAM through 2012. The first woman to become a member of the committee was
Sheila Widnall in 1983. Widnall went on to be the first female United States
Secretary of the Air Force (1993–1997). A second woman did not arrive on the
committee until 1996. However, in 2012 there were six women on the committee
of 35 (17 %) reflecting the increased number of women working in mechanics and
recognition of the fact that it was important to be a fully representative committee. As described previously, one of these women, Nadine Aubry, was the first
female Chair of the USNC/TAM for the 2006–2008 term. In 2012, the IUTAM
General Assembly elected Aubry to be the first female member of the IUTAM
Bureau.

Ethnic and Racial Diversity

The United States officially recognizes six ethnic and racial categories.
Wikipedia lists 57 contemporary ethnic groups. The names in the USNC/TAM
membership list show that there is a very wide variety of ethnic backgrounds
among the committee members. There is representation from North and South
America, Europe, and Asia. As to racial diversity, the first African-American to
join the committee was Lance Collins in 2006. Collins went on to become the
Chair of the Committee in 2010. A second African American, John Dabiri,
joined the committee in 2013.

Age Diversity

The average age of USNC/TAM committee members has been of concern for many
years. Senior members who are on the committee by virtue of their activities in
IUTAM weigh the average age to a higher number. The valuable input of these
members is particularly important at the international level. If the United States is to
play a leading role at the international level, there must be highly regarded individuals who have served in IUTAM for a significant number of years. If elected to the
IUTAM Bureau and then an office such as Treasurer, President (and then automatically Vice-President) this can result in 12 or more years of IUTAM service. As this
typically follows previous service in USNC/TAM, the individual would no longer
be "young."

IUTAM limits the number of Members-at-Large. The positions are limited to
individuals who have an international reputation for their scientific contributions,

and who have contributed to IUTAM over a period of years; thus, they are not "young." Younger members, both female and male, provide new ideas and energy to the committee. The committee has striven to identify rising young stars and get them involved in the committee. Member-at-large positions, society representatives, and IUTAM Congress Committee appointments are all methods used to involve new members.

The tables below show the distribution of ages for the 2011–2012 Committee, and the history of the average age of the committee from 2006 to 2012. For the 2011–2012 distribution, ages range from 36 to 87. The three oldest members, ages 87, 83, and 76, are all IUTAM Members-at-Large. Excluding these three members, the average drops down from 57.8 to 55.6.

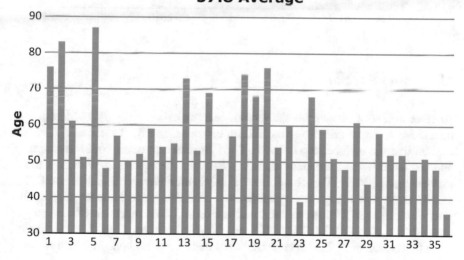

The history of the committee's average age from 2006 to 2012 shows that the efforts to bring in younger members had some positive results with the average lowered from 59.5 to 58.0 over the 6-year period. Since, typically, only three to four members are replaced on the committee each year (and the continuing members are all 1-year older) lowering the average age is not something that occurs at a rapid pace. As noted above, excluding the three IUTAM Members-at-Large for 2012 lowered the average age to 55.6.

The bottom line on all this is that while it is certainly desirable to bring in new, young, fresh ideas to the committee, it is also desirable to have the benefit of the expertise and experience of individuals with international reputations. It would be a mistake to overemphasize youth at the expense of international leadership.

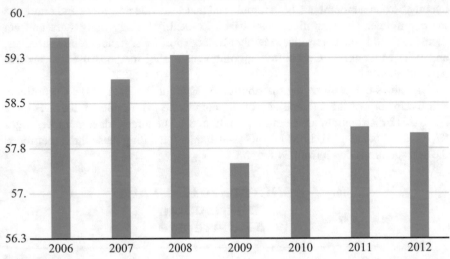

Mechanics Specialty

Another aspect of diversity is the area of mechanics specialty. IUTAM considers mechanics to consist of two areas, solid mechanics, and fluid mechanics. It is the custom that the membership of the IUTAM Bureau represents both areas equally, and that the IUTAM Presidency alternates between the two areas. With the natural, ever-changing world, the disciplines of mechanics have changed as well. The USNC/TAM now recognizes computational mechanics and biomechanics as distinct disciplines in addition to solid mechanics and fluid mechanics. The areas of specialty for 2011–2012 USNC/TAM membership includes 17 in solid mechanics, 13 in fluid mechanics, four in computational mechanics, and one in acoustics.

Chapter 15
USNC/TAM Activities

The USNC/TAM has been involved in a variety of activities over the years. As with most efforts of this type, some have been more successful than others. The major efforts are described in the following paragraphs.

Meetings with Governmental Agency Representatives

Minutes of committee meetings from as early as 1979 indicate interaction between committee members and representatives of government agencies such as NSF, ONR, and AFOSR. Minutes from later years show interactions with congressional staff members. It can be assumed that such interactions took place from the earliest days of the committee; however, the lack of minutes for meetings from the early days leaves us in the position of not being able to confirm such interactions between 1948 and 1979.

In 1979, Chairman Paul Naghdi reported that he and member Ronald Rivlin had visited NSF to ascertain how NSF had been reorganized as it relates to mechanics. In following years, in addition to frequent visits to governmental offices, it became the practice to invite various representatives of government agencies to attend USNC/TAM meetings when the meetings were held in Washington, D.C. These meetings were typically 2-day, Friday–Saturday affairs with the representatives giving presentations on Friday afternoon. The purpose of these presentations was for the representatives to present the status of their organization and financial conditions as they pertain to mechanics, and to give the committee members an opportunity to ask questions and make suggestions. All indications are that the sessions were highly valued by both the presenters and the committee members as a welcomed exchange of information and ideas.

© Springer International Publishing Switzerland 2016
C.T. Herakovich, *Mechanics IUTAM USNC/TAM*,
DOI 10.1007/978-3-319-32312-1_15

SCORDIM Reports

What does SCORDIM stand for? This has been a question, if not a running joke, for many years at USNC/TAM meetings. SCORDIM stands for *Sub Committee On Research Directions In Mechanics*. This definition is found in the minutes of a 1989 committee meeting. The story goes as follows. Norm Abramson (Chair 1986–1988) appointed a long-range planning subcommittee with Robert Plunket as the Chair and S. Antman, J. W. Dally and B. R. Noten as members. In 1988 Jim Dally presented the subcommittee's report to the full committee. The subcommittee recommended, among other things, "that the [USNC/TAM] Chair be authorized to develop a procedure for making periodic reviews of the recent trends and future research directions for Applied Mechanics." The report identified five principal areas of mechanics: (1) fluid mechanics, (2) sold mechanics, (3) computational mechanics, (4) dynamics and controls, and (5) experimental mechanics and materials. It was recommended that one or more reports be issued every year covering various aspects of one of the principal areas. With a 5-year cycle, each principal area would be revisited every 5 years.

By the 1989 meeting, Plunket and Dally were no longer members of the committee. However, Jim Dally (as a guest) was now leading the discussion along with Tinsley Oden. Dick Christensen, Chair of the USNC/TAM for 1988–1990, had refined the proposal suggesting that reports be issued once every 2 years rather than every year. Christensen asked Tinsley Oden to take the lead developing a plan to complete the first report on Computational Mechanics. As reported by Hodge in the May 1989 min, a lengthy discussion (1.5 h) was devoted to the proposal. In Hodge's summary of the discussion, he coined the name SCORDIM, but indicated that the first charge to the new committee would be to agree on an acceptable name. Apparently, the name was never changed as minutes of later meetings continued to refer to SCORDIM reports. The proposal for the SCORDIM reports was quite extensive including suggestions for organization, editors, editorial subcommittees, publication cycle, review procedure, costs, and funding.

During the discussion in 1989, Drucker is quoted as making a comment that showed far-reaching insight on the future of mechanics as an engineering science. Referring to the scope of the reports, he said ".. we should also be concerned with the value of mechanics as an engineering science. Mechanics can be very useful, but in a sense it's already been done by Newton. Mechanics seems to be disappearing as a label, and it's important to recognize this fact. Mechanics needs to be supported as a Science for the background of the future, not just to solve known problems. This is a hard point to make, but if we don't do it, who will? SCORDIM should balance this viewpoint about half and half with current applications."

The first report, on Computational Mechanics, was completed on schedule in 1990. However, it was not published until 1991. In a December 1991 letter from Oden to the USNC/TAM, he noted that it took over a year to have the report processed through the NRC. Minutes of the 1992 meeting indicate that the NRC charged the committee $32,000 to review the Computational Mechanics report.

This charge effectively took all the money in the committee's account, and according to committee members was spent without their authorization. Dana Caines, NRC Staff Associate, explained that approximately 25 % of the $32,000 was for converting the document submitted on a disk to WordPerfect and the rest was for the reviewing process. Committee members were very unhappy about the cost of working through the NRC to publish the report. Sid Leibovich, the current Chair, summarized the committee feelings saying that "the most charitable interpretation I can give of this incident is that it represents Gross Mismanagement." It was now obvious to the committee that they would have to find additional funding if they were to continue to publish similar SCORDIM reports through the NRC.

The second SCORDIM Report was published in 1996, 5 years after the first report. It was published by the American Institute of Physics, thereby eliminating the costly review process of the NRC. More than 1000 copies were either purchased from AIP or distributed freely to government agencies. The report was well received; however, concern was expressed over how long it took to be completed. The third SCORDIM report, on solid mechanics, was completed in 1999 and published in early 2000 as a special issue of the International Journal of Solids and Structures. As for the second report, the expense and delay associated with NRC review was avoided. There was no formal NRC review of either the 2nd or 3rd SCORDIM report.

During the 1998 meeting Tinsley Oden made the point that the original Computational Mechanics report was issued in 1990 and according to the original plan that each topic should be revisited every 10 years, it was time for the next volume on Computational Mechanics. He then volunteered to take the lead. Computational Mechanics II was published on schedule in January 2001. It was an 8-page, glossy, color, report updating the status and future directions for Computational Mechanics. The total cost of publication was the $1700.00 printing cost. There was no formal review by the NRC.

The success of Computational Mechanics II in terms of cost, time to publication, and the perceived improvement in style and length was apparent. It was felt that such a shorter publication had a better chance of being read by non-experts in the field, and a better chance of being written by people who were otherwise involved in more personal activities. Two reports of this type followed, one in 2006, on Fluid Dynamics and another in 2007 on Computational Mechanics and Composite Mechanics.

As of 2012, six SCORDIM reports have been issued by the USNC/TAM. Tinsley Oden emphasized that the reports should be written such that they were understandable to non-specialists such as high-level agency people. It is clear that the committee had trouble finding volunteers who would take the time to write such reports. The complete list of SCORDIM reports follows:

1991   Research Directions in Computational Mechanics (1991), by J. T. Oden (Ed.) 144 pages, National Academy Press.
1996   Research Trends in Fluid Dynamics (1996), by J L. Lumley, A. Acrivos, L. G. Leal, and S. Leibovich (Eds.), American Institute of Physics.

1999   Research Trends in Solid Mechanics (1999), by G. Dvorak (Ed). See: Int.
Journal of Solids and Structures, Vol. 37, pp 1–422, 2000.
2000   Research Directions in Computational Mechanics II (2000), by J. T. Oden,
T. Belytschko, I. Babuska and T. J. R. Hughes, J. Computational Methods Appl.,
Mechanics 8 pages.
2006   Research in Fluid Dynamics: Meeting National Needs, by J. Gollub, H.
Fernando, M. Gharib, J. Kim, S. Pope, A. Smits and H. Stone, 8 pages
2007   Research Directions in Computational and Composite Mechanics: Part 1:
Computational Mechanics by T. Belytschko, T. J. R. Hughes and N. Patankar,
Part 2: Mechanics of Composite Materials by C. T. Herakovich and C. E. Bakis,
8 pages.

U.S. Congresses of Applied Mechanics

The first U.S. Congress of Applied Mechanics was held June 11–16, 1951 at the
Illinois Institute of Technology in Chicago. This was 3 years after the committee
was officially formed in 1948. The second congress was held 3 years later in 1954
at the University of Michigan. A U.S. Congress of Applied Mechanics has been
held every 4 years since 1954. In 2002 and 2010, the congress was identified as the
U.S. Congress of Theoretical and Applied Mechanics as the international congresses had become to be named.

Each congress is hosted by a local organizing committee that typically is from
one university. The USNC/TAM selects the host for each congress based upon written proposals and oral presentations made by those seeking to host a congress.

Appendix J is a listing of all U.S. Congresses.

IUTAM Symposia and Summer Schools

Scientific symposia are sponsored by IUTAM to bring together a group of
active scientists in a well-defined field for the development of science within
that field. The number of participants is limited (recommended less than 100)
in order to achieve effective communication within the group. The number of
oral presentations is also limited to roughly eight per day. Participation is by
invitation only.

IUTAM typically sponsors no more than 16 symposia worldwide in any 2-year
period. Selection is based upon proposals submitted to IUTAM. The proposals are
reviewed and rated by the IUTAM solids or the fluids symposium panel. Final decisions as to the approved symposia are determined by vote of the entire IUTAM
General Assembly. The symposia are an excellent vehicle to introduce younger
researchers with those who are more established in the field.

IUTAM also sponsors a small number of Summer Schools for each 2-year period. The summer schools are intended to provide state-of-the-art lectures in new and emerging fields of mechanics. They are intended for researchers in mechanics, applied mathematics, physics, and engineering science.

The United States has been quite successful in having proposals for IUTAM symposia and summer schools approved by the General Assembly. Appendix K gives a listing of those held in the United States, or co-chaired by a person from the United States. There have been 47 symposia held in the United States, another ten co-chaired by the U.S. scholars and held outside the United States. There have been two Summer schools held in the United States.

Reviewing the list of symposia chairs shows that there have been many outstanding mechanics researchers who have participated actively in symposia, but have not been highlighted elsewhere in this history.

International Congresses of Applied Mechanics

International Congresses on mechanics have been held since 1924. Since 1926, they have been held at 4-year intervals except for a period during the Second World War. Prior to 1964 the organization of the International Congress of Applied Mechanics was supervised by the *International Committee for the Congresses of Applied Mechanics*. The organization of each congress was under the overall supervision of a Scientific Committee appointed by the International Committee. The organizational details of each congress were entrusted to a local Organizing Committee that also undertook the publication of the proceedings. Unfortunately, there is no central office from which proceedings may be ordered; for each volume, a request must be made to the publishers of that particular congress. A complete listing of congresses, including information of the proceedings, is available on the IUTAM website.

On September 4, 1964, the International Committee for the Congress of Applied Mechanics became a standing committee of IUTAM. However, this committee continues to have full responsibility for the selection of the host of IUTAM congresses and the scientific program. Members of the Congress Committee are elected by the full IUTAM General Assembly, but from that point forward, the Congress Committee acts as an independent body. Starting in 1964, the congresses were titled *International Congress on Theoretical and Applied Mechanics*.

Three International Congresses of Applied Mechanics have been held in the United States, the 5th Congress in 1938 at Cambridge, MA, jointly sponsored by Harvard University and the Massachusetts Institute of Technology, the 12th in 1968 at Stanford University and the 20th in 2000 in Chicago with the University of Illinois as the lead university for a consortium of 12 universities. Members of the USNC/TAM have been very active in most congresses presenting papers, organizing sessions, giving opening and closing lectures and receiving major awards. For each congress, the USNC/TAM reviews papers submitted from the United States and ranks them for consideration by the International Papers Committee.

Travel Grants to International Congresses

A major activity of the USNC/TAM has been securing funds from government agencies (NSF, AFOSR, ONR, and NASA) to support travel of researchers from the United States who are presenting papers at International Congresses. Special attention has been given to younger researchers in order to assist them in developing international exposure. The following table provides a summary of the grant activity beginning with 1976, the first year for which records are available. The table shows the year of the congress, location, total dollar amount of awards, and the number of grants awarded, when that number is available. There were no travel grants in 2000 when the International Congress was held in Chicago. As indicated in the table, the number of grants awarded has experienced a steady decrease over the years. This is the result of little to no increase in the available funds and the higher cost of travel to international locations. The number of applications for travel grants has always exceeded the available funds.

Travel Grant to International Congresses

Year	Location	Total	Grants	Year	Location	Total	Grants
1976	Delft	50,000	60	1996	Kyoto	60,000	44
1980	Toronto	30,000	70	2000	Chicago	0	0
1984	Lyngby	26,000	43	2004	Warsaw	45,000	36
1988	Grenoble	50,000	?	2008	Adelaide	66,000	33
1992	Haifa	21,600	?	2012	Beijing	44,000	22

International Prizes

IUTAM instituted two prestigious prizes in 2008. IUTAM awards these prizes only once every 4 years at the ICTAM congress. Three researchers from the United States have received these awards. The Rodney Hill Prize in Solid Mechanics was awarded to Michael Ortiz of Caltech in 2008 and Hujan Gao of Brown University received the Hill Prize in 2012. Howard Stone of Harvard University received the George Batchelor Prize in Fluid Mechanics in 2008.

USNC/TAM Operations Manual

In 1988, the Long Range Planning Committee chaired by Robert Plunkett recommended the development of a manual outlining the functions of USNC/TAM officers and committees. Secretary Philip Hodge took the lead in this endeavor.

Hodge, other officers and subcommittee Chairs all wrote an operations section for their respective position. The final Operations Manual is very detailed and runs to 34 pages in a Microsoft Word file. Because of its length and the fact that it has been updated a number of times as committee operations changed, it will not be included in this history. It should be available from the National Academies.

Chapter 16
Finances

The total 1948 financial support for the USNC/TAM was $400 for IUTAM dues, paid through contributions from the seven participating societies (ASME, SESA, ASCE, AIChE, APS, IAS, and AMS). No financial records for the years 1948–1976 are available. Thus, there is no record of how the IUTAM dues were paid nor who covered the travel expenses during those years. It is likely that the members of the committee secured funds for their travel from their home institutions or paid them from personal funds or research grants. It is also likely that, after 1968, the NAS assisted with funds for IUTAM dues and travel. There is some evidence that the U.S. State Department paid dues to international organizations in earlier periods.

The minutes of the January 1979 committee meeting provide the first detailed financial accounting for the committee. This report shows a total budget of $32,846 provided by NSF for the 2-year period November 1, 1976 through December 31, 1978, essentially $16,423 per year. These funds covered travel to USNC/TAM and IUTAM General Assembly meetings, IUTAM Dues, NAS staff support and NAS overhead charges. An additional NAS General and Administrative fee of 11 % on all expenses is present in later financial reports. NAS and the NSF agreed to this fee arrangement.

The available records show that the National Science Foundation has provided funding for the committee, through the National Academy of Sciences, for all years since 1979. However, the detailed record of this funding is spotty at best. It is fair to say that NAS has maintained tight control over what data they have been willing to share with the committee. NAS policy on financial matters has varied over the years, in part due to the committee's alignment within the academy structure. In 2 years (1985–1986 and 1986–1987), the committee was given no financial information and was told that, as a policy, financial data was not to be

© Springer International Publishing Switzerland 2016
C.T. Herakovich, *Mechanics IUTAM USNC/TAM*,
DOI 10.1007/978-3-319-32312-1_16

made available to the committee. In other years, general financial data was available, but the NAS said that the committee could not see staff salary data. In contrast, the 3-year proposal to NSF for April 1, 1990 to March 31, 1993 provided complete information on the support staff salary and percent time devoted to the committee. The Chapter on Relations with the NAS discusses this point in more detail.

There are two categories of committee funds, NAS funds, and private funds. The NAS funds are those provided by the NSF. The private funds are those accumulated by the committee as return from the U.S. National Congresses on Applied Mechanics. The NAS has complete control over the NSF-provided funds. The committee generally believes that it has complete control over the private funds, but, in fact, once the committee agreed to be associated with the NAS, it ceded control of those funds to the NAS as well.

NSF Funding

The NSF provides funds to the committee based upon proposals submitted by the NAS. The proposals are written by the NAS with little or no input from the USNC/TAM. This is particularly true as it relates to financial data. It appears that the system used by the NAS has varied over the years. At times, the NSF funds were granted to the NAS as a block grant and then NAS decided how to distribute the funds between their administrative offices and the individual committees. In later years, individual proposals were made to the NSF on behalf of specific NAS committees; officers of the USNC/TAM were given the opportunity to review the text of these proposals, but they were not permitted to be involved, in any manner, with the funding requested.

The chart below provides the (known) history of the yearly funding by NSF, through the NAS. For the years that no confirmed data is available, the chart is set at zero, which was clearly not the case. It must be noted that two variations in the data are: (1) in early years it was often the case that IUTAM dues were paid out of a different account, and (2) at times (on 4-year intervals) the NSF portions of travel funds to International Congresses were included in the committee budget whereas in other years, these funds were in a totally separate proposal to NSF. The chart shows that funding for the committee increased in a relatively smooth manner until the 2006–2007 year. Significant percentage increases were evident in 2006–2007 and again in 2010–2011. These large increases can be explained in part as a result of the fact that a new administration was in place at BISO (Board on International Scientific Organizations), the organizational body within NAS that USNC/TAM was under. BISO controlled the budget. The percentage of BISO expenditures for staff, indirect expenses, overhead, and the general and administrative fee also increased significantly during this period. They were 46% from 2006 to 2010 and 51% from 2010 to 2013.

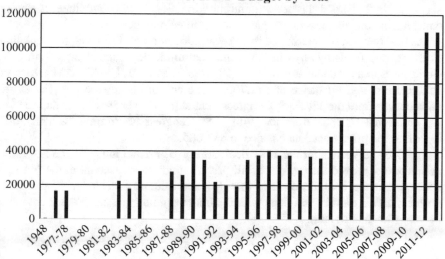

USNC/TAM Private Funds

The USNC/TAM accumulates *private funds* as proceeds from U.S. National Congresses that convene every 4 years. One exception was that in 2009, the account received an infusion of $5464 when the ACDM (Association of Chairmen of Department of Mechanics) dissolved and they transferred their balance to the USNC/TAM private account. In early years, the private funds in an account at the NAS generated interest income. In later years, NAS discontinued the interest provision on private accounts. Although called private, by virtue of being a committee of the NAS, the expenditure of these funds is subject to final NAS authority.

The committee uses the private funds as needed when funds are not available under the NSF support. The committee uses the private funds to assist with ICTAM travel, publication of SCORDIM reports and, during 1 year, 2008, $2500 in additional financial assistance for each successful IUTAM symposia proposed for the United States. The total financial assistance was then $7500 for each successful symposium as IUATM provides $5000 in support of every IUTAM Symposia. Three U.S. symposia received this assistance.

The chart below shows the (known) history of the private fund balances. The 1979 financial report indicated a private fund balance of $17,255 as of Nov. 1, 1976 and $20,009 as of Dec. 31, 1978. The 1979 minutes indicated that the 6th National Congress (1974 at Boulder) provided $4200 in return to the committee. During the period 1993–2001, minutes indicate that the committee received no reports as to the private account, most probably because there were no funds in the private account. The NAS completely exhausted the USNC/TAM private funds when they charged

$32,000 to review the first SCORDIM report, using all of the funds in the private account. NAS charged this fee without authorization of the committee; it caused considerable unhappiness on the part of committee members.

As best that can be determined from committee minutes, following the depletion of the private funds by NAS in 1991, the committee deposited returns from U.S. Congresses in an account at Iowa State University. In 2001, USNC/TAM Treasurer Jim Hill reported a balance of $11,700 in the account at Iowa State. The entire balance was from the 13th U.S. Congress held at the University of Florida in 1998. An additional $21,385 from the 14th U.S. Congress at Virginia Tech in 2002 increased the private account balance to $33,085.

In 2003, 12 years after the 1991 depletion of all private funds by the NAS, the committee approved the transfer of all private funds to an account at the NAS. It appears that the committee made this decision without knowledge of the 1991 problems. This transfer was done with the understanding that any NAS fees would be very low (5 % or less).

USNC/TAM Private Funds

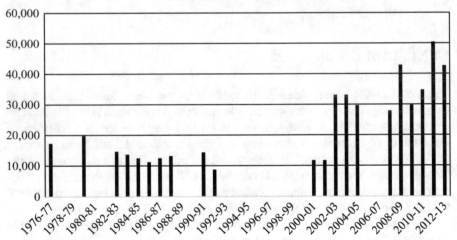

Chapter 17
Relations with the National Academy of Sciences

A history of the USNC/TAM is incomplete without a review of its relationship with the National Academy of Sciences/National Research Council. As will be demonstrated in the following, at times, the relationship has been less than the desired optimum.

The U.S. engineering scientists who specialized in mechanics organized a group of like-minded scientists in 1948 in order to have a formal organization for participation in the International Congress of Applied Mechanics. They also organized the U.S. National Congresses of Applied Mechanics first held in 1951 and quadrennially since 1954. These engineering scientists acted solely on behalf of themselves and their colleagues. The individual members, and the seven societies that these men were actively involved with, approved the original charter in 1948. Unfortunately, a copy of the original charter has not been located. As established, the committee was independent and not affiliated directly with any society or governmental organization. Committee decisions were not subject to review by any other body. ASME did provide secretarial assistance and a place to meet during the early years. In 1949, the USNC/TAM became the adhering U.S. body to IUTAM.

In 1958, the National Academy Sciences invited the USNC/TAM to become a committee of the NAS. Nick Hoff was chair of the committee at the time. The committee did not accept this invitation until 1965, officially becoming a committee of the Division of Physical Sciences of the National Academy of Sciences in 1966. The Chapter USNC/TAM in this book provides details surrounding this invitation. When the USNC/TAM accepted the invitation to join the NAS, it ceded the final authority over the committee to the NAS, and the NAS became the official adhering U.S. body to IUTAM. What little history is known for the early years of the committee was obtained from the files of F. N. Frenkiel (secretary of the committee from 1970 to 1982), a limited selection of records from the NAS for 1966, and a few fragments of comments from committee minutes.

A detailed review of the available minutes (meetings in 1966, 1976, and 1979 forward) and personal experience as a member of the committee for 18 years (beginning in 1996), including 12 years as secretary, provided the writer with insight

© Springer International Publishing Switzerland 2016
C.T. Herakovich, *Mechanics IUTAM USNC/TAM*,
DOI 10.1007/978-3-319-32312-1_17

into the relationship. Prior to joining the NAS, the USNC/TAM functioned as an independent committee. In a 1985 committee meeting Nick Hoff made the statement "Before 1967 USNC/TAM was independent of NAS. We were promised lots of support if we joined the other USNCs." This was in response to a statement by a representative from the NAS Bureau of International Affairs, that the Office of International Affairs "supports only the international part" of the USNC/TAM work.

After agreeing to join the NAS in 1966, USNC/TAM became a committee under the umbrella of the NAS and subject to its policies, procedures, and governance. The committee surrendered its independence. Since that time, the NAS expanded in size, modified its organizational structure a number of times and exerted more control over the USNC/TAM.

The organizational structure of the NAS evolved over the years; it is now one of three organizations within The National Academies. The other two organizations are the NAE (National Academy of Engineering, 1964) and the IOM (Institute of Medicine, 1970). In 1958, when the NAS invited the USNC/TAM to join the NAS, neither the NAE nor the IOM existed. As the National Academies grew over the years, the USNC/TAM became a much smaller cog in the more complex organizational structure of the National Academies.

The figure below shows a partial organizational chart of the National Academies as of 2013. The National Research Council (NRC) is the operating arm of the National Academies. The NRC is organized around six major divisions; Behavioral and Social Sciences and Education (BSSE), Earth and Life Studies (ELS), Engineering and Physical Sciences (EPS), Policy and Global Affairs (PGA), Institute of Medicine (MED in the chart), and Transportation Research Board (TRANS). A seventh program unit is the Gulf Research Program. The USNC/TAM (one of twenty-some such committees) is located further down the organizational chart as a committee under the Board on International Scientific Organizations (BISO) which is one of the five Boards in the Policy and Global Affairs Division (PGA). The PGA Division is also home to eleven other programs and committees. Obviously, as of 2013, the USNC/TAM is a very small cog in a very big wheel.

The USNC/TAM was aligned under the (new) Board on International Scientific Organizations (BISO) and the Policy and Global Affairs Division when the NRC reorganized in 2001. The mission of the Policy and Global Affairs Division is "to help improve public policy, understanding, and education in matters of science, technology, and health with regard to national strategies and resources, global affairs, workforce, and the economy." The mission of BISO is to "strengthen science for the benefit of society through U.S. leadership, collaboration, and representation in international scientific organizations and initiatives." In summary, the USNC/TAM is under two organizational levels whose missions are concerned with international policy and global affairs as opposed to the development and dissemination of new science. This observation is further strengthened by the fact that, as of this writing, the Director of BISO and the Executive Director of PGA both are lawyers with essentially no scientific or engineering background.

Since the USNC/TAM is under PGA and BISO, one might assume that international policy and global affairs are a major activity of the USNC/TAM. In studying the available minutes and personally being aware of what has transpired since 1996, there is not a single incidence where the committee was involved in an international policy or global affairs activity. The committee emphasis is, and has been, on the details of scientific progress. The same is true for IUTAM. It has had very little activity that is of the international policy or global affairs variety.

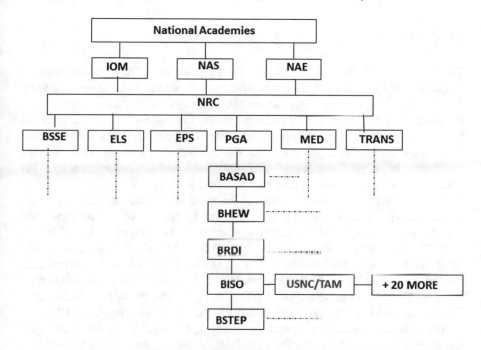

All of the above points to the fact that the activities of the USNC/TAM are some-what inconsistent with the mission of BISO. The fundamental interests of the USNC/TAM are at variance with the typical NAS committee. Further, unlike other committees, the USNC/TAM has a history of governing itself.

The committee history shows that there are issues related to its relationship with the NAS/NRC that have caused concerns for committee members. At times, the concerns have been so strong that committee members have raised the question "why is it necessary for our committee to be associated with the NAS/NRC?"

The following paragraphs detail interactions between the committee and the NAS for three distinct periods, 1976–1997, 1997–2008, and 2008–2012.

1976–1997

The minutes of meetings during these years indicate frequent, lengthy discussions about issues with the NAS. The issues dealt with four primary topics: constitution, financial, committee size, and length of member service. Between 1976 and 1997, the committee was aligned with several different organizations within the NAS structure; at least six different NAS support staff worked with the USNC/TAM during this period. These frequent organizational and staff changes resulted in an inconsistent message to the committee from the NAS.

Constitutional Issues

The October 1979 minutes refer to a proposal from the NRC Board of Internal Operations and Programs (BIOP) that officers of the committee "be nominated by the Committee but that actual appointment be made by the Chairman of the National Research Council rather than by elections by the Committee." A second proposal would limit the total service of a regular member to 12 years. Dan Drucker reported that "he was a member of BIOP and that the constitutional changes were a recommendation by a staff committee of BIOP; they are not requirements of the NRC. There is no sentiment in BIOP for uniformity."

This same issue arose in some detail in the minutes of March 21, 1980. A report of the Ad Hoc Committee Regarding Proposed Changes in the By-Laws and Constitution indicated a strong negative reaction to the changes proposed by the NRC. During the 1980 discussion, Naghdi, Hoff and Frenkiel commented on the origins of the relationship between the USNC/TAM and the NAS-NRC. They pointed out that the committee had been independent since 1948 and it was the U.S. Adhering Body to IUTAM since 1949. During that time, it was the only U.S. National Committee associated with an ICSU Union not connected with the NAS/NRC.

The 1980 differences were eventually resolved, but similar constitutional issues were raised again in later years. The NAS/NRC wanted all committees to have near identical constitutions and by-laws. The USNC/TAM was an outlier primarily because it wanted to follow the policies and procedures it had established when it was independent from the NAS. One example is that the committee had been functioning for several years with a member designated as Treasurer. When the committee proposed a change in its constitution to reflect this, the National Academies responded that the committee could not have a Treasurer; only the National Academies could have a Treasurer.

Committee Size

The size of the committee has been a long-standing issue between the NAS and USNC/TAM. This appears to be a fundamental difference between what the NAS views as the role of their committees, i.e., limited to international policy and global affairs, and the mission of the USNC/TAM as stated in its 1971 constitution. The USNC/TAM has always had a twofold mission, participation, and leadership in the international mechanics community (primarily through IUTAM), and promotion of mechanics within the United States. When the NAS invited USNC/TAM to become a committee of the NAS in 1958, the Committee was already larger than the six to eight members envisioned by the NAS. The membership of USNC/TAM continued to grow (33 members in 2013) as it endeavored to meet its mission of promoting mechanics in the United States and recognizing U.S. members of IUTAM as members of the committee.

When reporting on a strategic planning subcommittee in 1995, Tinsley Oden noted that the committee is performing three functions outside those that the NRC proposed as a common scope for all U.S. National Committees:

(a) The U.S. National Congress
(b) Liaison with program directors of funding agencies
(c) SCORDIM reports

Clearly, there is a larger mission for the USNC/TAM than the NAS envisions for U.S. National Committees. This difference is the result of the fact that the committee was independent before joining the NAS and had been performing the functions outside the NAS common scope prior to joining the NAS. For example, the first U.S. National Congress on Theoretical and Applied Mechanics was in 1951 at the Illinois Institute of Technology. The U.S. Congresses have been held quadrennially ever since 1954. The NAS, PGA, and BISO do not have a mission or interest in national congresses.

Length of Service

The length of service on the USNC/TAM continues to be an issue because NAS wants to have a 12-year guideline. However, the USNC/TAM realizes that in order to fill its desired role in IUTAM, it is necessary for its members to have long-term experience in IUTAM if there is hope that one of its members might be elected to the IUTAM Bureau. Once elected to the Bureau, the strong possibility also exists that the member will become an officer and eventually, the President of IUTAM. This can easily add twelve or more years as the person advances through the offices (4 year terms) up to President and Vice-President (automatic after serving as President). In addition, other members serve on important IUTAM committees including the Congress Committee, the Fluids Symposium Panel, and the Solids Symposium Panel. These appointments are typically 8 years each and IUTAM experience is desirable prior to appointment. If elected as the Chair of the Congress Committee or one of the Panels, this can add eight more years of service. Finally, the historical knowledge of members who have served for a longer period is invaluable both from the standpoint of IUTAM and USNC/TAM. This has added importance in view of the fact that the NAS staff-support person typically has a tenure of fewer than 5 years. Hence, there is very little historical carryover at the NAS.

A somewhat celebrated case, referred to as the *Acrivos Problem*, was the proposed appointment of Andy Acrivos in 1988. The APS nominated him to be their representative to the committee for 1987–1991. Acrivos previously had served the committee in several different capacities, including Chair. His appointment as the APS representative was not in conflict with the USNC/TAM Constitution and By-Laws, which the NRC had approved. Nonetheless, the appointment was denied by the Chair of the Board on Physics and Astronomy (which the committee was under at the time) because of a "Policy of the National Research Council against excessively long-term appointment." Further, the Board Chair notified the APS directly without consultation with the committee.

The situation was further amusing as in 1988 the USNC/TAM was moved from the Board on Physics and Astronomy to the Committee on Engineering and Technical Systems (CETS). CETS had no firm policy on length of service so, after the fact, the issue became moot. In 2004, IUTAM elected Acrivos as a Member-at-Large; he thereby became a member of the USNC/TAM again and served many additional years. In 2001, the President of the United States awarded the National Medal of Science to Acrivos.

It is most interesting that the NAS/NRC is most interested in the international activities of the committee and yet does not appreciate the fact that long-term participation on the committee is necessary if the U.S. is to have the desired leadership positions in IUTAM.

Financial

The period 1976–1997 was not a particularly good time for the committee regarding financials and related matters. The committee often found that it did not have sufficient funds to cover travel and IUTAM dues. It was often the case that the NAS staff did not provide a financial report at annual meetings. In 1986 when the committee expressed unhappiness over the lack of a financial report, they were told "such internal financial information was not for release." The most significant problem may have been when the NAS depleted the entire $32,000 balance of private funds to cover charges for reviewing the first SCORDIM report in 1991. The NAS/NRC did this without the approval or knowledge of the committee.

The committee *private funds* are those received as return from National Congresses. They are distinguished from those provided by federal agencies, primarily NSF. The private funds were held by the NAS during some periods, and at other times, the private funds were held in a bank account outside the NAS. The expenditure of these funds when held by the NAS has been, at times, a problem for the committee.

The dire financial condition in 1994 caused Chair Tinsley Oden to appoint a Crisis Subcommittee. The subcommittee reviewed the manner in which the committee functioned and suggested recommendations for change with special attention to financial matters. The subcommittee used, as a starting point, the report that Oden had written to the NAS/NRC in response to a questionnaire on the U.S. National Committees.

In that report, Oden stated that there was "a general feeling among members of our Committee that more flexibility is needed in the way the Committee manages its own budget within the NRC." He stated further, "Many of our members feel that our current relationship with the NRC has been the source of unnecessary difficulties in pursuing our agenda."

The full committee discussion of the situation centered on two main topics, where should the USNC/TAM be located within the NAS/NRC structure in order to have the desired support and flexibility, and what can the committee do about the dire financial situation. Comments by some committee members during this discussion were indicative of considerable unhappiness with the NRC.

The committee unhappiness was due, at least in part, to the fact that the committee had accepted the NAS policy that any funds received by committees of the NAS were to be "solicited, received, and distributed in accordance with prevailing policies and practices of the National Academies."

1997–2008

In the 1997 NAS structure, the USNC/TAM was under the National Materials Advisory Board (NMAB), which was under the Committee on Engineering and Technical Systems, which was under the Committee on International Organizations and Programs (CIOP), which was under the NRC. The Office of International Affairs (OIA), within CIOP, was the primary contact for the committee. Wendy White was the Director of the International Programs office and Lois Peterson was the program officer from CIOP working with the USNC/TAM, under Wendy White. At the June 1997 meeting, Peterson made a presentation summarizing the NAS organizational structure, the role of USNCs, and the relationship of the NAS with ICSU (International Council of Scientific Unions). Peterson made two important statements during her presentation. One, the USNCs "are creations and instrumentations of the Academy and responsible to it." Clearly, this has been the view of the NAS, but in fact, the USNC/TAM was not a *creation* of the NAS. The USNC/TAM was an independent committee for 18 years prior to accepting the *invitation* to become a committee of the NAS. Peterson also stated that the USNC/TAM was the only committee representing engineering and that, in many ways, the USNC/TAM is as responsible to the NAE as it is to the NAS. A review of the minutes and personal observations do not indicate any meaningful interaction between the USNC/TAM and the NAE.

The belief, or attitude, within the NAS that the USNC/TAM is a *creation* of the NAS may be the fundamental cause of problems between the USNC/TAM and the NAS.

Notwithstanding the above statement about the creation of the committee, the years working with Peterson as the staff officer were quite harmonious. With Peterson's leadership, the NAS appeared to do everything within their power to accommodate the committee's wishes. During the 1997 meeting, Bob Schafrik,

Director of the NMAB, made the statement that White and Peterson "have taken a very pragmatic approach to the issue on the constitution." Following events proved that this was true. Indeed, they took a very pragmatic approach on all issues when working with the committee.

In 1998, the NAS reorganized once again and USNC/TAM was put under the Division for International Organizations and Academy Cooperation (IOAC). In 2001, another reorganization at the NAS put the USNC/TAM under BISO (Board on International Scientific Organizations) with Wendy White as the BISO Director and Lois Peterson continuing as the Program Officer.

Throughout her tenure as the Program Officer, Peterson did an excellent job on all aspects of the various interactions with the committee. Peterson submitted detailed reports on NSF funding, committee private funds, and international travel funds on a regular basis. Working with the USNC/TAM Executive Committee, Peterson oversaw the transfer of the private funds from Iowa State University to an account with the NAS. She reported that the BISO policy for private funds in an account at NRC would incur a range of fees (2.6–18.6%), based upon the type of expenditure, and that no interest would be generated. On several occasions, she expressed interest in having more interaction between the USNC/TAM and the NAE.

When the committee submitted a revised constitution and by-laws for approval, Peterson reported that the General Counsel would not allow the committee to have a Treasurer. She worked with the Executive Committee to reach a compromise that was acceptable to the General Counsel.

With the approval of the USNC/TAM executive committee, Peterson and the Committee Secretary developed a plan whereby each society in USNC/TAM would submit the names of two qualified individuals as nominees to be their representative on the committee. The USNC/TAM executive committee would then select one of these individuals as the representative, taking into consideration scientific qualifications and committee diversity. This system worked quite well.

One of the final activities that Peterson was involved with for the committee was the travel grant program for the 2008 International Congress in Adelaide Australia. The program included, for the first time, a breakfast with the grantees on the first morning of the congress and assignment of a committee member as mentor to each of the young grantees. Peterson accomplished this work from her office in Washington.

2008–2012

In 2007, Wendy White, the long time Director of BISO retired. At the 2007 meeting of the USNC/TAM, Lois Peterson announced that she would be stepping down as the liaison person for the committee. The NAS/NRC appointed a new Director of BISO who appointed a new Program Officer (liaison person) for the USNC/TAM. The new Director of BISO had only joined the NAS/NRC in 2005.

The new Program Officer, who also was new to the NAS/NRC, lasted only a few months as the liaison. The committee had three different liaisons between 2008 and 2010. All three were young internationals having been in the United States for a limited time; they lacked experience with the NAS/NRC and the local scientific community. Further, neither the BISO Director nor the Executive Director of PGA had any background in mechanics, science, or engineering; they both were lawyers. Clearly, there was very little understanding of the historical significance of the USNC/TAM or the field of mechanics by those in positions of authority over the USNC/TAM.

Coincident with these staff changes there was a noticeable change in the attitude of the staff. The attitude changed from one of administrative support to one of being in charge. The following paragraphs describe some noteworthy items.

NSF Funding

The preparation of budgets for proposals to NSF has always been the sole responsibility of the NAS/NRC. When committee officers indicated a desire to participate in budget preparations, the response from BISO was along the lines, *such financial details are not open to the public.* However, prior to submission of the proposal to the NSF, BISO requested that the officers of the USNC/TAM appear at the NSF with the BISO liaison person to provide support for the proposal. The discussions at NSF were on the technical merits of the proposal, not the financial aspects. Thus, the officers were in the position of promoting a budget they had not developed nor seen. Their colleagues in the mechanics community who would review the proposal on behalf of the NSF would of course see all the fine details of the proposal. The reviewers would know more about the budget than the USNC/TAM officers would. As indicated in a previous chart, the budgets for the committee experienced step function increases in 2006 and again in 2010.

Private Funds

In 2008, the USNC/TAM decided to use some of its *private funds* to supplement IUTAM awards for the U.S.-sponsored IUTAM symposia. The idea was to provide more incentives for the U.S. proposers. A USNC/TAM subcommittee reviewed the U.S. proposals prior to submission to IUTAM, an IUTAM subcommittee then rated them and, finally, the IUTAM General Assembly voted on all proposals. Several U.S. proposals were approved by IUTAM. When BISO was advised as to who should receive the funds, BISO required the symposia chairs to write another proposal to the NAS/NRC. When, after a considerable delay, the funds were distributed, the letters of transmittal stated that the grant was from the NAS/NRC. The letter did not mention the USNC/TAM as the source of the funds.

Staff Travel to International Meetings

The IUTAM General Assembly meets every 2 years at locations around the world. In the years when there is an International Congress on Theoretical and Applied Mechanics (every 4 years), the General Assembly meets during the Congress. In the off-congress years, the General Assembly meets for 2 days, most often in the home country of the current IUTAM President.

After the new BISO Director took over in 2007, the Director made the decision to send the staff liaison to all meetings of the IUTAM General Assembly. This of course required considerable funding. Money spent sending a staff person to a meeting where they knew nothing about the subjects under discussion meant that there was less money for committee operations. In addition, in view of the short tenure of each liaison, there was very little carryover. The only previous time when the liaison attended a General Assembly meeting was the 2000 meeting in Chicago. The meeting was not at an international location and the USNC/TAM held its annual meeting in Chicago, the day before the Congress.

When I pointed out that we already had nine people attending the 2010 General Assembly meeting in Paris and that we would submit a report, it was not necessary to send the liaison, the Director responded it was no problem, "*NSF was paying for it.*" The liaison traveled to Paris for a 2-day meeting that he knew nothing about. The Director also sent this liaison to the IUTAM Congress in Beijing (a 5-day scientific congress) where the General Assembly met in conjunction with the Congress. In this case, there can be no question that the money would have been spent better sending a young researcher to Beijing so they could gain international experience.

During that ICTAM congress in Beijing, another U.S. Committee (Astronomical) was holding a 2-week congress at the same convention center. I met the BISO Director (who had no background in Astronomy) in a hotel elevator; she explained that she was there for the full 2-weeks to oversee several social functions. Thus, two members of BISO were at the same convention center, at the same time, to oversee social functions that were (or could have been) arranged previously from the United States.

A final note on this travel policy is that the USNC/TAM liaison person who attended the Paris and Beijing meetings was no longer with BISO as of October 2013; hence, no carryover.

Administrative Support

The administrative support from the NAS/NRC/BISO has been a constantly changing story. Both the organizational entity that USNC/TAM was a part of changed often over the years, and the individual who provided the liaison to the committee changed often. Eleven different liaisons served between 1976 and 2012, on average, a new liaison every 3.2 years.

When the Operations Manual for the committee was written in April 1991, one section was entitled *Duties of the Staff Associate*. As stated, the duties were to support and assist the committee with meeting logistics, preparing and reporting cost estimates and drafting proposals, maintaining liaison with policy makers, and assisting officers with the execution of their responsibilities. The duties listed were administrative—not authoritative.

When Lois Peterson (the NAS staff officer) updated that section of the manual in 2001, she added a section indicating that the staff officer would "Assist in identifying potential committee members, assemble background information on candidates, and prepare nominations and appointment packages. Ensure that NRC (as the NAS was referred to) committee records are up-to-date." As secretary at the time, I worked closely with Lois on these matters and I felt that the working relationship was excellent.

By 2008, this attitude of staff support had changed to more of one of staff control. And, while the BISO Director stated on several occasions that their office could do more to assist the committee, two attempts to have a BISO staff member take the minutes of committee meetings were unsuccessful.

USNC/TAM Website

When websites were introduced around 2000, there was a reasonably good working relationship between the committee and liaison Peterson. The committee was able to have listed on the website, those items the committee thought desirable and the overall structure of the website clearly indicated that it was a USNC/TAM website. When the new Director of BISO was appointed in 2007, this all changed. BISO made the decision that websites of all committees had to be the same. The new design made the websites look like a BISO website rather than a USNC/TAM website. In fact, one mechanician who wanted to serve on the committee sent an email asking if he could "become a member of BISO."

When I suggested to the BISO Director that there was a problem with the USNC/TAM website, she asked me to send suggestions for change. When I did this, she responded that what I was suggesting was not possible because all websites had to be the same.

New Member Selection

The committee takes great pains to select new members who had a strong record of accomplishment in mechanics as well as meet the needs of the committee for balance in all of the various diversity considerations, e.g., solid mechanics, fluid mechanics, computational mechanics, merging areas of mechanics, gender, age, and ethnic background. The policy introduced in 2006 requesting two nominees from each society significantly improved the committee diversity while maintaining the high quality. The committee had a subcommittee that reviewed the nominees and made a recommendation to the full committee. In essentially all cases, someone on the committee knew the nominees personally. This knowledge was vital when one thinks about the role that might eventually be played by the member in IUTAM as well as in the USNC/TAM.

After the change in BISO Directorship in 2007, considerable pressure was put on the committee to change the process for selection of new committee members. It was suggested strongly that BISO contact the various societies and individuals for appointments. This was presented as a time saver for the committee. Yet at the same time, BISO was saying that they needed more funding from the NSF to support the committee. In addition, if BISO took over this role, the invitations to join the committee would be coming from someone who had no background in mechanics and would be unknown to the invitee. It would have seemed like an invitation from BISO rather than the USNC/TAM. The committee rejected this approach.

Another aspect of new member selection was approval by the NAS. Prior to 2008, the committee considered approval pro forma. After 2008, the BISO Director made it a point to remind the committee that we were not the final authority when selecting new members. New members had to be approved and appointed by the President of the NAS—which was really a way of saying that appointments had to be approved by BISO and the PGA where there also was no expertise in mechanics. When Ralph Cicerone, the President of the NAS, met with the committee in 2012, his opening comment to the committee was "I really don't know anything about this committee." That was not a surprising statement. He was not a mechanician. It was an honest statement. Just another example of the disconnect between what the committee was being told by BISO and what was actuality.

2012 Strategic Planning

In 2011, Lance Collins Chair of the USNC/TAM, appointed a nine-member Strategic Planning Subcommittee. The subcommittee's purpose was to identify specific activities that bring the greatest benefit to the mechanics community. The subcommittee met a number of times, by conference call, over the fall and winter of 2011–2012. They also met with representatives of the NSF to discuss the committee's financial situation. The subcommittee submitted their report to the full committee at the 2012 annual meeting. The planning subcommittee addressed three fundamental questions:

- Who are we?
- What do we do?
- How do we do it?

Following discussion, the full committee adopted the following recommendations:

1. Society representatives serve one 4-year term and rotate off, with exceptions made either for outstanding service or for USNC/TAM or IUTAM officers;
2. Whenever possible, nominate IUTAM members as USNC/TAM members-at-large.
3. New members to the USNC/TAM committee will be asked to take responsibility for initiating one SCORDIM Report upon joining the committee.
4. Whenever possible, SCORDIM reports will be published in an appropriate peer-review journal. The Publications Committee will assist in developing relationships with key journals that publish short topical reviews (e.g., MRS Bulletin).
5. Establish a Liaison Subcommittee that would be available to:

   (a) Advise the agencies on research trends and directions.
   (b) Recommend members of advisory boards.
   (c) Assist in the staffing of program directors of these agencies.
   (d) Assist in developing strong proposal reviewer pools for specific programs.
   (e) Help disseminate information from these agencies back to the technical societies and the research community at large.
   (f) Provide support as needed to sensitize Congress to the importance of research to the country's well-being.

6. Improve relationship with the National Academies:

   (a) The Chair, Vice-Chair, and Secretary of USNC/TAM will serve as Co-PIs on all future proposals to federal funding agencies, including the NSF.
   (b) A memorandum of understanding that clearly defines expectations for committee officers and the National Academies staff will be written and signed by both parties.
   (c) Private funds that formerly were held in a private account, that then were transferred to an account at the National Academies, will be returned to a private 501c3 account that will be maintained by the Secretary of USNC/TAM.
   (d) Work with the National Academies staff to discuss ways to develop and maintain a more vibrant and up-to-date USNC/TAM website, preferably with the USNC/TAM logo in the most prominent position on the page.

Epilogue

The engineering science of mechanics has had a long and glorious history providing society with the fundamental laws, written as mathematical equations, that govern the behavior of solids and fluids—when at rest or in motion. Mechanics also includes the experimental verification of the mathematical equations. The field has developed slowly, but confidently, over the course of 3000 years. The contributions made by the engineers and scientists working in the field have enhanced the lives of their fellow men and women through a continual progression of advancements.

The individuals who have made the most significant contributions to the field are true giants of society. Through their intellect and confidence, they have made bold statements that were often in conflict with the beliefs that preceded their contributions. Although often in competition in their attempts to solve a particular problem, they have maintained exceptional personal relationships typically sharing their ideas with their competitors. Their competitors were also colleagues. This is demonstrated most dramatically by the activities and relationships within IUTAM where engineering scientists from throughout the world share their ideas and friendships.

The USNC/TAM is a committee of very successful individuals in the United States engineering science community. Many have been honored throughout the world for their accomplishments. They are hardworking, honest people with the highest degree of integrity. They know their field well and are fiercely proud of it. They have a desire to improve the daily life of their fellow man.

It has been an honor and a privilege to work alongside so many outstanding individuals over the course of more than 50 years; to have become friends with many of them has been icing on the cake.

© Springer International Publishing Switzerland 2016
C.T. Herakovich, *Mechanics IUTAM USNC/TAM*,
DOI 10.1007/978-3-319-32312-1

Appendix A: IUTAM Bureau Members

© Springer International Publishing Switzerland 2016
C.T. Herakovich, *Mechanics IUTAM USNC/TAM*,
DOI 10.1007/978-3-319-32312-1

Year	President	Vice President	Treasurer	Secretary	Member	Member	Member	Member
2012	Tvergaard DEN	Pedley UK	Eberhard GER	Dias FRA	Aubry USA	Rubin ISR	Schrefler ITA	Yang CHN
2010-11	Pedley UK	Freund USA	Olhoff DEN	Dias FRA	Chernousko RUS	Gupta IND	Lund CHI	Thess GER
2008-09	Pedley UK	Freund USA	Olhoff DEN	Dias FRA	Chernousko RUS	Goldhirsch ISR	Gupta IND	Thess GER
2004-07	Freund USA	Moffatt UK	Engelbrecht EST	van Campen NED	Kambe JPN	Kluwick AUT	Olhoff DEN	Zheng CHN
2000-03	Moffatt UK	Schiehlen GER	Freund USA	van Campen NED	Cercignani ITA	Engelbrecht EST	Narasimha IND	Salencon FRA
1996-99	Schiehlen GER	Wijngaarden NED	Freund USA	Hayes IRL	Engelbrecht EST	Moffatt UK	Tatsumi JPN	Wang CHN
1992-95	Wijngaarden NED	Germain FRA	Boley USA	Ziegler AUT	Chernyi RUS	Moffatt UK	Schiehlen GER	Tatsumi JPN
1988-91	Germain FRA	Lighthill UK	Wijngaarden NED	Schiehlen GER	Boley USA	Chernyi RUS	Imai JPN	Ziegler AUT
1984-87	Lighthill UK	Drucker USA	Wijngaarden NED	Schiehlen GER	Germain FRA	Hult SWE	Imai JPN	Ishlinski RUS
1980-83	Drucker USA	Niordson DEN	Becker GER	Hult SWE	Germain FRA	Lighthill UK	Sedov RUS	Tani JPN
1976-79	Niordson DEN	Gortler GER	Drucker USA	Hult SWE	Germain FRA	Lighthill UK	Olszak POL	Sedov RUS
1972-75	Gortler GER	Koiter NED	Drucker USA	Niordson DEN	Legendre FRA	Lighthill UK	Olszak POL	Sedov RUS
1968-71	Koiter NED	Roy FRA	Gortler GER	Niordson DEN	Hoff USA	Lighthill UK	Olszak POL	Sedov RUS
1967	Roy FRA	Temple UK	Koiter NED	Gortler GER	Hoff USA	Mikhailov RUS	Olszak POL	Sedov RUS
1964-66	Roy FRA	Temple UK	Koiter NED	Gortler GER	Hoff USA	Olszak POL	Parkus AUT	Sedov RUS
1960-63	Temple UK	Odqvist SWE	Koiter NED	Roy FRA	Hoff USA	Gortler GER	Muskhelishvili RUS	Ziegler AUT
1956-59	Odqvist SWE	Dryden USA	Temple UK	Roy FRA	Ackeret SUI	Colonnetti ITA	Grammel GER	Koiter NED
1952-55	Dryden USA	Peres FRA	Temple UK	den Dungen BEL	Burgers NED	Odquist SWE	Grammel GER	Goldstein ISR
1948-51	Peres FRA	Southwell UK	Dryden USA	Burgers NED	Colonnetti ITA	den Dungen BEL	Favre SUI	Solberg NOR

Appendix B: USNC/TAM Constitution

Purpose

The U.S. National Committee on Theoretical and Applied Mechanics

(a) Will promote theoretical and applied mechanics in the United States.
(b) Will strive to maintain a balance between the various established subfields of mechanics and to accommodate and include emerging subfields.
(c) Will represent the United States in the International Union of Theoretical and Applied Mechanics, on behalf of the National Academy of Sciences.

Membership

The U.S. National Committee shall be made up of the members listed in II(a) through II(f) below, each of whom shall serve until his or her successor is duly appointed. An individual who is qualified under more than one paragraph below shall be considered one member only. Unless specifically stated otherwise, all members shall be voting members. Terms will begin on November 1 and end on October 31.

(a) Representatives from each of the societies listed below shall be nominated by the societies and appointed by the Chair of the National Research Council for a 4-year term, with approximately one-fourth of their terms expiring each year. Terms are renewable once, but representatives are eligible for membership under the other articles below.

Acoustical Society of America
American Academy of Mechanics
American Institute of Aeronautics and Astronautics
American Institute of Chemical Engineers
American Mathematical Society

© Springer International Publishing Switzerland 2016
C.T. Herakovich, *Mechanics IUTAM USNC/TAM*,
DOI 10.1007/978-3-319-32312-1

American Physical Society (Division of Fluid Dynamics)
American Society for Testing and Materials
American Society of Civil Engineers (Engineering Mechanics Division)
American Society of Mechanical Engineers (Applied Mechanics Division)
Society for Experimental Mechanics
Society of Engineering Science
Society of Industrial and Applied Mathematics
Society of Naval Architects and Marine Engineers
Society of Rheology
U.S. Association of Computational Mechanics

(b) Members-at-large shall be nominated by the U.S. National Committee on Theoretical and Applied Mechanics and appointed by the Chair of the National Research Council for a 3-year term, with the terms of approximately one-third expiring each year. At least four, but no more than eight such members shall serve on the committee at the same time.

(c) Officers of the U.S. National Committee shall be members of the committee as specified in Article III.

(d) The General Chair of a scheduled U.S. National Congress shall be an ex-officio member for one 4-year term that begins approximately two and one-half years before and ends approximately one and one-half after the Congress.

(e) Voting members of the International Union of Theoretical and Applied Mechanics General Assembly, Congress Committee, and Symposium Panels, who are residents of the United States of America, shall be ex-officio members. Their terms shall be for the duration of their service in the named office. This includes the U.S. delegates to the General Assembly who shall be nominated by the committee and appointed by the Chair of the National Research Council for 2-year terms.

(f) The Chair of the National Research Council Division on Engineering and Physical Sciences, the Chair of the National Research Council National Materials Advisory Board, and the Foreign Secretary of the National Academy of Sciences shall be non-voting ex-officio members. In addition, the National Materials Advisory Board may appoint a liaison.

Officers

The Chair, Vice-Chair, and Secretary shall be nominated by the committee and appointed by the Chair of the National Research Council for concurrent 2-year terms. The Chair shall be nominated from among current committee members and shall serve one term only as Chair, but the most recent available Past Chair shall continue to serve as an officer for an additional 2-year term. The Secretary and Vice-Chair need not be members of the committee before their appointments; there is no statutory limit to the number of terms they may serve.

Finances

All funds (e.g., contributions, registration fees, contracts, and grants) for support of the USNC/TAM and its programs shall be solicited, received, and disbursed in accordance with prevailing policies and practices of the National Academies.

It is the National Academies' responsibility to assure payment of the annual subscriptions to the International Union of Theoretical and Applied Mechanics. The USNC/TAM shall keep the National Academies informed of any proposed changes in subscription rates and seek their agreement on such changes.

Amendments

The Constitution may be amended by an affirmative vote of two-thirds of the members of the U.S. National Committee entitled to vote, provided that 3-week notice of the proposed amendments has been given to all members of the committee and provided that the amendments receive subsequent approval of the Governing Board of the National Research Council.

- Adopted June 13, 1958
- Approved by participating Societies March 25, 1959
- Amended July 15, 1965
- Ratified by the Governing Board of the National Research Council April 24, 1966
- Amended by the Governing Board December 7, 1969
- Amended by the Governing Board September 26, 1971
- Amended by the Governing Board June 1980
- Amended by the Governing Board September 1985
- Amended by the Governing Board April 1993
- Amended by the Governing Board November 2002
- Approved by Policy and Global Affairs Executive Director on behalf of Governing Board, October 11, 2005
- Approved: Policy and Global Affairs Executive Director on behalf of Governing Board, April 22, 2008

Appendix C: USNC/TAM By-Laws

U.S. National Committee on Theoretical and Applied Mechanics

I. Participating societies

(a) The participating societies referred to in Article I of the Constitution shall consist of the following bodies:

- American Academy of Mechanics
- American Society of Civil Engineers (Engineering Mechanics Division)
- American Society of Mechanical Engineers (Applied Mechanics Division)
- American Mathematical Society
- American Institute of Chemical Engineers
- American Physical Society (Division of Fluid Dynamics)
- American Institute of Aeronautics and Astronautics
- Society for Experimental Mechanics
- American Society for Testing and Materials
- Society of Rheology
- Society of Engineering Science
- Society of Industrial and Applied Mathematics
- Society of Naval Architects and Marine Engineers
- Acoustical Society of America
- U.S. Association for Computational Mechanics

(b) Additional bodies may be added to the list of participating societies within the limits specified in Article IIa of the Constitution. Any such addition will require an affirmative action by two-thirds of the voting members of the U.S. National Committee.

(c) A body may be voluntarily removed from the list of participating societies upon a written request by an appropriate officer of the participating society.

(d) A body may be involuntarily removed from the list of participating societies only by the following procedure:

© Springer International Publishing Switzerland 2016
C.T. Herakovich, *Mechanics IUTAM USNC/TAM*,
DOI 10.1007/978-3-319-32312-1

(1) A motion to request removal of a body may be made at any regularly scheduled meeting of the U.S. National Committee. Such a motion must include
the reasons for removal, and must be approved by two-thirds of the voting
members of the U.S. National Committee.

(2) Copies of the motion will be sent immediately to an appropriate officer of
the participating society and to its representative, and they shall each be
given 90 days to file a request to remain on the list of participating societies
and to respond to the reasons stated in the motion.

(3) At the next regularly scheduled meeting of the U.S. National Committee
after the expiration of the 90-day period referred to in the preceding paragraph, the Secretary shall make a motion to remove the body in question
from the list. The discussion of this motion shall include the original motion
to request removal, and any responses received. If the motion is approved
by two-thirds of the voting members of the U.S. National Committee, it
shall be effective immediately; otherwise the request for removal shall be
inoperative and the matter may not be reopened at the current meeting.

II. Members

(a) The terms of the representatives nominated by the bodies listed in Article
I(a) above shall be 4 years commencing on November first; the terms of
approximately one-fourth of them shall expire on October 31 of each year
or when their successors are duly elected. These representatives can be
nominated for no more than two consecutive 4-year terms but are eligible
for membership under Articles II(b) through II(f) of the Constitution. Each
society shall nominate two qualified candidates to be their representative.
The USNC/TAM Executive Committee shall select one of the two for recommendation to The National Academies as the society representative.
Appointment is made by The National Academies.

(b) Under Article II(b) of the Constitution the number of members-at-large
shall be between 4 and 8, each serving a 3-year term commencing on
November first; the terms of approximately one-third of them shall expire
on October 31 of each year. Members-at-large are elected by the USNC/
TAM and appointed by The National Academies.

(c) The General Chair of a U.S. National Congress shall be an ex-officio member from the time the Congress is scheduled through October 31 approximately one and one-half years after the Congress; he or she shall become a
voting member starting November 1 approximately two and one-half years
before the Congress.

(d) The members included in Article II(e) of the Constitution include five
Delegates elected by the Committee. At each annual meeting of the committee in an odd numbered year, these delegates shall be elected by a majority
vote of the U.S. National Committee from among its members at the time of
the election. Each election or re-election shall be for a 2-year term starting
November 1 of the year of the election and ending on October 31 of the next
odd-numbered year or when their successors are duly elected. The terms of

all ex-officio members named in Article II(d) of the Constitution shall be for the duration of their service in the named office.

(e) The Officers included in Article III of the Constitution shall be elected by the committee at its Annual Meeting. The Chair shall be elected from among the voting members. He or she may not serve more than one consecutive term as Chair, but is eligible for election to a different office. The Past chair is the most recent Chair. The Secretary need not be a member of the U.S. National Committee prior to his or her election; there is no statutory limit to the number of terms served by the Secretary.

(f) The Foreign Secretary of the National Academy of Sciences, the Chair of the Division on Engineering and Physical Sciences, and the Chair of the National Materials and Manufacturing Board shall be non-voting ex-officio members of the committee. In addition, the committee may invite one additional member of the National Materials and Manufacturing Board to be an ex-officio voting member.

(g) If a position becomes vacant for any reason, it shall be treated as follows: A representative's position shall be nominated by the appropriate Society as soon as possible to serve the remainder of the term. A Member-at-Large position shall remain vacant until the next meeting at which time a new election shall be held for the remainder of the term. An officer or a U.S. member of the General Assembly of the International Union of Theoretical and Applied Mechanics shall be replaced by consensus of the USNC/TAM Executive Committee to serve for the remainder of the term.

(h) If, for any reason, an individual serves less than 2 years of a term that service shall not be counted for purposes of eligibility for reappointment or re-election. If 2 or more years are served, that service shall count as service for a full term.

III. Executive committee

The Executive Committee shall consist of the four Officers plus at least two other members appointed by the Chair. It shall be empowered to act for the Committee between meetings of the Committee, but any actions taken may be nullified by majority vote of those members present at the next meeting.

IV. Operations

(a) An Annual Meeting shall be held once in each calendar year. Additional meetings may be scheduled. Business of the Committee may also be transacted by email or by telephone conference.

(b) A simple majority of the members of the Category II(a) of the Constitution shall constitute a quorum.

(c) The Chair may appoint Subcommittees to assist in carrying out the business of the U.S. National Committee. All Subcommittees shall report to the full Committee at each Annual Meeting.

V. Voting

   (a) Unless otherwise specified, all motions made at a meeting shall be decided
       by a simple majority of the votes cast. In case of a tie vote, the motion shall
       fail.
   (b) Any action taken by the Executive Committee must have the concurrence of
       at least 75 % of the Executive Committee.
   (c) The Executive Committee may authorize the Secretary to place any motion
       before the membership by means of a web or email ballot. Any such ballot
       must contain a deadline for its return that is at least 3 weeks distant from the
       date the ballot is sent. In order to become effective, any such motion must
       receive an aye vote from a majority of the voting members of the Committee.
       A motion to amend the Constitution or Articles I or V of the by-laws must
       receive an aye vote from at least two-thirds of the voting members of the
       Committee. The Secretary shall tabulate the results of the ballot including
       the number of aye and nay votes. This information shall be immediately
       given to the members via the same medium as the ballot, and shall also be
       reported at the next meeting and recorded in the Minutes thereof.
   (d) The Executive Committee may authorize the Secretary or a staff member to
       arrange a telephone conference (see item III).
   (e) If a representative is unable to attend a meeting, the Society may send an
       alternate instead. The alternate must present to the Secretary a signed (email
       is sufficient) authorization from the representative or from an officer of that
       Society before any vote in which he or she participates. The alternate shall
       have all the rights and privileges of the original representative, except that an
       alternate may vote on an amendment to the Constitution or the by-laws only
       if he or she has a directed proxy from the representative for that vote.
   (f) If any member is unable to attend a meeting, he or she may give a proxy to
       any other member to vote on all issues that come before the meeting. To be
       effective, a signed (email is sufficient) copy of the proxy must be delivered
       to the Secretary before the first vote on which it is to be used. A proxy may
       be totally free or it may be partially or fully directed.

VI. Amendments of the by-laws

   (a) The by-Laws except for I and V may be amended by an affirmative vote of
       a simple majority of those members of the full Committee entitled to vote,
       provided that 3 weeks notice of the proposed amendments has been given
       to all members of the Committee and provided that the amendments are not
       in conflict with the Constitution.
   (b) By-Laws I (Participating Societies) and VI (Amendments of the by-Laws)
       may be amended by an affirmative vote of two-thirds of those members of
       the full Committee entitled to vote, provided that 3 weeks notice of the

proposed amendments has been given to all members of the Committee and provided that the amendments are not in conflict with the Constitution.

Adapted June 13, 1958.
Amended October, 1985;
May 1990;
April 1993;
May 1993;
May 1996;
April 28, 2001.

Appendix D: Societies Represented in USNC/TAM

Following is a listing of the member societies and the year they were founded.

Acoustical Society of America (ASA)

Established in 1929.

American Academy of Mechanics (AAM)

Established in 1942.

American Institute of Aeronautics & Astronautics (AIAA)

Established in January 1963 as AIAA, it was the merger of the American Rocket Society (begun in 1930 as the American Interplanetary Society) and the Institute of the Aerospace Sciences (established in 1932 as the Institute of Aeronautical Sciences).

American Institute of Chemical Engineers (AIChE)

Founded in 1908.

American Mathematical Society (AMS)

Founded in 1888.

American Physical Society (APS)

Society founded in 1899, Division of Fluid Dynamics established in 1948.

ASTM International (formerly American Society for Testing and Materials)

ASTM founded in 1898. In 2001, name changed to ASTM International.

American Society of Civil Engineers (ASCE)

Founded in 1852. Engineering Mechanics Division founded in 1950. In 2007, Engineering Mechanics Division replaced by the Engineering Mechanics Institute.

© Springer International Publishing Switzerland 2016
C.T. Herakovich, *Mechanics IUTAM USNC/TAM*,
DOI 10.1007/978-3-319-32312-1

American Society of Mechanical Engineers (ASME)

Founded in 1880, the Applied Mechanics Division established in 1927.

Society for Experimental Mechanics (SEM)

Founded in 1943 as The Society for Experimental Stress Analysis. Name changed to Society for Experimental Mechanics ~1990.

Society for Industrial & Applied Mathematics (SIAM)

Incorporated in 1952.

Society of Engineering Science (SES)

Founded in 1963.

Society of Naval Architects & Marine Engineers (SNAME)

Organized in 1893.

Society of Rheology (SOR)

Formed in 1929.

U.S. Association for Computational Mechanics (USACM)

Incorporated in 1988.

Appendix E: USNC/TAM Membership

Appendix E.1: USNC/TAM Membership by Year

For 1948–1982, the committee membership is based upon the IUTAM Annual Reports. Ex-officio members are indicated (EO). The committee had 12 voting members and one ex-official member in 1948. This number grew over the years because the number of societies represented increased from 7 to 15, and in 1969, the committee adopted a policy that members of the IUTAM Congress Committee are members of USNC/TAM. The committee had 35 members in 2012. The following list is a best-effort attempt. Clearly, the list is not 100 % accurate for many years prior to 1982.

1948	Members-at-Large: H. L. Dryden (Chair), Th. von Kármán, S. P. Timoshenko, J. C. Hunsaker, R. von Mises. Society Representatives: H. W. Emmons (ASME), R. D. Mindlin (SESA), M. G. Salvadori (ASCE), T. B. Drew (AIChE), R. J. Seeger (APS), N. J. Hoff (IAS), E. Reissner (AMS), and C. E. Davies (ex-officio from ASME, serving as secretary)
1949	Members-at-Large: H. L. Dryden (Chair), Th. von Kármán, S. P. Timoshenko, J. C. Hunsaker, R. von Mises. Society Representatives: H. W. Emmons (ASME), R. D. Mindlin (SESA), M. G. Salvadori (ASCE), T. B. Drew (AIChE), R. J. Seeger (APS), N. J. Hoff (IAS), E. Reissner (AMS), and C. E. Davies (ex-officio from ASME, serving as secretary)
1950–1957	H. L. Dryden, H. W. Emmons, N. J. Hoff, J. C. Hunsaker, Th. von Kármán, Y. H. Ku, R. D. Mindlin, R. von Mises, M. G. Salvadori, R. J. Seeger, S. P. Timoshenko
Note: Ku came to the United States and MIT (from China) in 1950	
1952–1954	H. L. Dryden, J. C. Hunsaker, N. J. Hoff, A. T. Ippen, Th. von Kármán, Y. M. Ku, M. H. Martin, R. D. Mindlin, R. von Mises, N. M. Newmark, S. P. Timoshenko
1954–1956	J. M. Burgers, H. L. Dryden, S. Goldstein, N. J. Hoff, J. C. Hunsaker, A. T. Ippen, Th. von Kármán, Y. M. Ku, R. D. Mindlin, R. von Mises, N. M. Newmark, S. P. Timoshenko

(continued)

© Springer International Publishing Switzerland 2016
C.T. Herakovich, *Mechanics IUTAM USNC/TAM*,
DOI 10.1007/978-3-319-32312-1

(continued)

Note: Sydney Goldstein relocated from Manchester (by way of the Technion) to Harvard in 1955	
Note: Jan Burgers relocated from Delft to Maryland in 1955	
1956–1958	J. M. Burgers, H. L. Dryden, F. N. Frenkiel, S. Goldstein, N. J. Hoff, M. Hetenyi, Th. von Kármán, Y. M. Ku, W. Prager, W. Ramberg, S. P. Timoshenko
1958–1960	J. M. Burgers, H. L. Dryden, F. N. Frenkiel, S. Goldstein, M. Hetenyi, N. J. Hoff, Y. M. Ku, W. Prager, W. Ramberg, Th. von Kármán, S. P. Timoshenko
1960–1962	J. M. Burgers, H. L. Dryden, F. N. Frenkiel, S. Goldstein, M. Hetenyi, N. J. Hoff, Th. von Kármán, Y. H. Ku, W. Prager, W. Ramberg, S. P. Timoshenko
1962–1964	J. M. Burgers, D. C. Drucker, H. L. Dryden, S. Goldstein, F. N. Frenkiel, M. Hetenyi, N. J. Hoff, Th. von Kármán, Y. H. Ku, E. P. Popov, W. Prager, W. Ramberg, S. P. Timoshenko
Note: Th. von Kármán died on May 6, 1963	
1965	N. J. Hoff, D. C. Drucker, F. N. Frenkiel, W. Prager, G. F. Carrier, J. M. Burgers, H. L. Dryden, S. Goldstein, J. P. den Hartog, E. P. Popov, S. P. Timoshenko, Y. H. Ku
Note: 1965 was the first year that IUTAM listed members of the Congress Committee in the Annual Reports. When appropriate, they are included as members of the USNC/TAM from here forward	
Note: Dryden died on Dec. 2, 1965	
1966–1968	J. M. Burgers, G. F. Carrier, D. C. Drucker, F. N. Frenkiel, S. Goldstein, J. P. den Hartog, N. J. Hoff, Y. H. Ku, E. P. Popov, W. Prager, S. P. Timoshenko
1969	A. Acrivos, R. H. Bing, J. M. Burgers, G. F. Carrier, S. H. Crandall, T. J. Dolan, D. C. Drucker, F. N. Frenkiel, S. Goldstein, N. J. Hoff, J. A. Hutcheson, J. B. Keller, Y. H. Ku, E. H. Lee, H. W. Liepmann, E. P. Popov, W. Prager, H. Rouse, O. B. Schier (EO), G. B. Schubauer, R. Smoluchowski, S. P. Timoshenko
As detailed in a 1969 letter from George F. Carrier (Chairman) to the National Research Council	
1964–1970	B. A. Boley, J. M. Burgers, G. F. Carrier, D. C. Drucker, F. N. Frenkiel, S. Goldstein, J. P. den Hartog, N. J. Hoff, J. B. Keller, Y. H. Ku, E. H. Lee, H. W. Liepmann, E. P. Popov, W. Prager, S. P. Timoshenko
1970–1972	B. A. Boley, J. M. Burgers, G. F. Carrier, D. C. Drucker, F. N. Frenkiel, S. Goldstein, N. J. Hoff, J. B. Keller, P. S. Klebanoff, Y. H. Ku, E. H. Lee, H. W. Liepmann, E. P. Popov, W. Prager, S. P. Timoshenko
Note: Timoshenko died on May 29, 1972	
1972–1974	J. D. Achenbach, B. A. Boley, B. Budiansky, J. M. Burgers, J. D. Cole, S. H. Crandall, D. C. Drucker, S. Goldstein, F. N. Frenkiel, G. H. Handelman, N. J. Hoff, D. J. Joseph, J. B. Keller, P. S. Klebanoff, Y. H. Ku, E. H. Lee, H. W. Liepmann, P. C. Paris, E. P. Popov, J. R. Rice
1974–1976	J. D. Achenbach, B. A. Boley, B. Budiansky, J. M. Burgers, J. D. Cole, S. H. Crandall, D. C. Drucker, F. N. Frenkiel, S. Goldstein, G. H. Handelman, N. J. Hoff, D. J. Joseph, J. B. Keller, Y. H. Ku, E. H. Lee, M. V. Morkovin, P. C. Paris, J. R. Rice, R. Rivlin
1976	J. C. Achenbach, M. L. Baron, B. A. Boley, H. Brenner, B. Budiansky, J. M. Burgers, J. D. Cole, S. H. Crandall, J. W. Dally, D. C. Drucker, F. Essenburg, F. N. Frenkiel, S. Goldstein, G. H. Handelman, N. J. Hoff, J. B. Keller, Y. H. Ku, E. H. Lee, M. V. Morkovin, P. M. Naghdi, P. C. Paris, J. R. Rice, R. Rivlin, J. L. Sackman, P. G. Saffman, W. R. Schowalter

(continued)

USNC/TAM Meeting: May 14, 1976, Washington, DC	
USNC/TAM Meeting: July 20, 1976, location unknown	
1976–1978	H. N. Abramson, J. C. Achenbach, M. L. Baron, B. A. Boley, H. Brenner, B. Budiansky, J. M. Burgers, J. D. Cole, S. Crandall, J. W. Dally, D. C. Drucker, F. Essenburg, F. N. Frenkiel, S. Goldstein, G. H. Handelman, N. J. Hoff, J. B. Keller, Y. H. Ku, H. Markovitz, P. M. Naghdi, P. C. Paris, R. S. Rivlin, J. L. Sackman, P. G. Saffman, W. R. Schowalter
1978–1979	H. N. Abramson, M. L. Baron, H. Brenner, B. A. Boley, B. Budiansky, J. M. Burgers, R. M. Christensen, J. D. Cole, J. W. Dally, S. H. Davis, D. C. Drucker, F. N. Frenkiel, S. Goldstein, G. H. Handelman, T. J. Hanratty, N. J. Hoff, L. N. Howard, J. B. Keller, Y. H. Ku, H. Markovitz, J. Miklowitz, J. W. Miles, E. M. Murmann, P. M. Naghdi, P. C. Paris, R. Peskin, R. S. Rivlin, A. Roshko, W. R. Schowalter, R. A. Toupin
1979	H. N. Abramson, M. L. Baron, H. Brenner, B. A. Boley, B. Budiansky, J. M. Burgers, R. M. Christensen, J. D. Cole, J. W. Dally, S. H. Davis, D. C. Drucker, F. N. Frenkiel, S. Goldstein, G. H. Handelman, T. J. Hanratty, N. J. Hoff, L. N. Howard, J. B. Keller, Y. H. Ku, H. Markovitz, J. Miklowitz, J. W. Miles, E. M. Murmann, P. M. Naghdi, P. C. Paris, R. Peskin, R. S. Rivlin, A. Roshko, W. R. Schowalter, R. A. Toupin
USNC/TAM Meeting: January 19, 1979, Washington, DC	
USNC/TAM Meeting: October 30, 1979, Boston, MA	
1980	H. N. Abramson, M. L. Baron, B. A. Boley, B. Budiansky, J. M. Burgers, R. M. Christensen, J. D. Cole, J. W. Dally, S. H. Davis, D. C. Drucker, F. N. Frenkiel, S. Goldstein, T. J. Hanratty, N. J. Hoff, L. N. Howard, J. B. Keller, Y. H. Ku, H. Markovitz, J. Miklowitz, J. W. Miles, P. M. Naghdi, P. C. Paris, R. L. Peskin, R. S. Rivlin, A. Roshko, W. R. Schowalter, R. A. Toupin
USNC/TAM Meeting: March 21, 1980, Pasadena, CA	
USNC/TAM Meeting: August 18, 1980, Toronto, Canada	
1981	H. N. Abramson, A. Acrivos, S. S. Antman, B. A. Boley, J. M. Burgers, M. M. Carroll, R. M. Christensen, J. D. Cole, S. Corrsin, J. W. Dally, S. H. Davis, D. C. Drucker, F. N. Frenkiel, S. Goldstein, T. J. Hanratty, N. J. Hoff, J. B. Keller, Y. H. Ku, H. Markovitz, P. M. Naghdi, Y. H. Pao, P. C. Paris, R. L. Peskin, R. S. Rivlin, J. L. Sackman, R. T. Shield, R. Skalak
USNC/TAM Meeting: November 19, 1981, Washington, DC	
Note: Burgers died on June 7, 1981	
From 1982 onward, committee membership is from the minutes of the UNSC/TAM meetings	
1982	H. N. Abramson, J. A. Acrivos, S. S. Antman, B. A. Boley, M. M. Carroll, R. M. Christensen, J. D. Cole, R. D. Cooper, S. Corrsin, J. W. Dally, S. H. Davis, D. C. Drucker, F. N. Frenkiel, S. Goldstein, T. J. Hanratty, P. G. Hodge, Jr., N. J. Hoff, J. W. Hutchinson, J. B. Keller, Y. H. Ku, H. Markovitz, P. M. Naghdi, Y. H. Pao, R. L. Peskin, R. S. Rivlin, J. L. Sackman, R. Skalak, R. T. Shield
USNC/TAM Meeting: June 21 & 22, 1982, Ithaca, NY	
1983	H. N. Abramson, A. Acrivos, S. S. Antman, E. B. Becker, B. A. Boley, M. M. Carroll, R. M. Christensen, R. D. Cooper, S. Corrsin, J. W. Dally, S. H. Davis, R. Di Prima, D. C. Drucker, F. N. Frenkiel, S. Goldstein, T. J. Hanratty, P. G. Hodge, Jr., N. J. Hoff, J. W. Hutchinson, J. B. Keller, Y. H. Ku, P. A. Libby, H. Markovitz, P. M. Naghdi, Y. H. Pao, J. L. Sackman, R. T. Shield, R. Skalak, C. E. Taylor

(continued)

(continued)

USNC/TAM Meeting: April 21, 1983, Washington, DC	
USNC/TAM Meeting: November 17, 1983, Boston, MA	
1984	H. N. Abramson, A. Acrivos, S. S. Antman, E. B. Becker, B. A. Boley, M. M. Carroll, R. M. Christensen, R. D. Cooper, S. Corrsin, J. W. Dally, S. H. Davis, R. Di Prima, D. C. Drucker, F. N. Frenkiel, J. Goddard, S. Goldstein, T. J. Hanratty, P. G. Hodge, N. J. Hoff, J. W. Hutchinson, J. B. Keller, Y. H. Ku, P. A. Libby, H. Markovitz, P. M. Naghdi, R. Nordgren, Y. H. Pao, R. Plunkett, J. L. Sackman, R. T. Shield, R. Skalak, C. E. Taylor, S. Widnall
USNC/TAM Meeting: April 27, 1984, Washington, DC	
1985	H. N. Abramson, A. Acrivos, S. S. Antman, E. B. Becker, B. A. Boley, R. M. Christensen, R. D. Cooper, S. Corrsin, J. W. Dally, S. H. Davis, D. C. Drucker, F. N. Frenkiel, J. Goddard, S. Goldstein, T. J. Hanratty, P. G. Hodge, Jr., N. J. Hoff, J. W. Hutchinson, J. B. Keller, Y. H. Ku, P. A. Libby, R. Nordgren, R. Plunkett, J. L. Sackman, R. T. Shield, C. E. Taylor, M. Tulin, S. Widnall
USNC/TAM Meeting: October 4, 1985, Washington, DC	
1986	H. N. Abramson, A. Acrivos, S. S. Antman, T. Belytschko, B. A. Boley, R. M. Christensen, R. D. Cooper, J. W. Dally, S. H. Davis, D. C. Drucker, L. B. Freund, J. Goddard, S. Goldstein, T. J. Hanratty, P. G. Hodge, N. J. Hoff, C. S. Hsu, J. W. Hutchinson, J. B. Keller, Y. H. Ku, J. P. Lamb, P. A. Libby, R. Nordgren, R. Plunkett, R. T. Shield, C. E. Taylor, M. Tulin, S. Widnall
USNC/TAM Meeting: June 15, 1986, Austin, TX	
1987	H. N. Abramson, J. D. Achenbach, A. Acrivos, S. S. Antman, T. Belytschko, B. A. Boley, R. S. Brodkey, C. F. Chen, R. M. Christensen, R. D. Cooper, J. W. Dally, D. C. Drucker, D. Frederick, L. B. Freund, J. Goddard, S. Goldstein, P. G. Hodge, Jr., N. J. Hoff, C. S. Hsu, J. W. Hutchinson, Y. H. Ku, J. P. Lamb, B. R. Noton, J. T. Oden, R. Plunkett, C. E. Taylor, M. Tulin, S. Widnall
USNC/TAM Meeting: May 10, 1987, Washington, DC	
1988	H. N. Abramson, J. D. Achenbach, A. Acrivos, S. S. Antman, T. Belytschko, B. A. Boley, R. S. Brodkey, C. F. Chen, R. M. Christensen, R. D. Cooper, C. F. Dafermos, J. W. Dally, D. C. Drucker, D. Frederick, L. B. Freund, T. L. Geers, J. Goddard, P. G. Hodge, N. J. Hoff, C. S. Hsu, J. W. Hutchinson, Y. H. Ku, J. P. Lamb, S. Leibovich, B. R. Noton, J. T. Oden, R. Plunkett, C. E. Taylor, M. Tulin, M. Van Dyke, S. Widnall
Note: Goldstein died on 22 January 1989	
USNC/TAM Meeting: April 17, 1988, Washington, DC	
1989	H. N. Abramson, J. D. Achenbach, A. Acrivos, T. Belytschko, D. Bogy, B. A. Boley, R. S. Brodkey, C. F. Chen, R. M. Christensen, R. D. Cooper, C. F. Dafermos, C. Dalton, D. C. Drucker, D. Frederick, L. B. Freund, T. L. Geers, J. Goddard, P. G. Hodge, Jr., N. J. Hoff, C. S. Hsu, J. W. Hutchinson, D. Joseph, Y. H. Ku, J. P. Lamb, S. Leibovich, B. R. Noton, J. T. Oden, C. E. Taylor, M. Tulin, M. Van Dyke, R. P. Wei
USNC/TAM Meeting: May 5 & 6, 1989, Washington, DC	
1990	H. N. Abramson, J. D. Achenbach, A. Acrivos, D. Bogy, B. A. Boley, R. S. Brodkey, C. F. Chen, R. M. Christensen, C. F. Dafermos, C. Dalton, D. C. Drucker, M. Fourney, D. Frederick, L. B. Freund, T. L. Geers, J. Goddard, P. G. Hodge, N. J. Hoff, D. D. Joseph, Y. H. Ku, L. G. Leal, S. Leibovich, B. R. Noton, J. T. Oden, P. Spanos, C. E. Taylor, M. Tulin, W. Vorus, R. P. Wei

(continued)

(continued)

USNC/TAM Meeting: May 20, 1990, Tucson, AZ	
1991	J. D. Achenbach, A. Acrivos, T. Belytschko, D. Bogy, B. A. Boley, R. S. Brodkey, C. F. Chen, R. M. Christensen, S. Crandall, C. F. Dafermos, C. Dalton, D. C. Drucker, M. Fourney, D. Frederick, L. B. Freund, T. L. Geers, J. Goddard, P. G. Hodge, N. J. Hoff, D. D. Joseph, A. Kobayashi, Y. H. Ku, L. G. Leal, S. Leibovich, J. T. Oden, P. Spanos, M. Tulin, W. Vorus, R. P. Wei

USNC/TAM Meeting: May 3 & 4, 1991, Washington, DC	
1992	J. D. Achenbach, A. Acrivos, D. Bogy, B. A. Boley, R. S. Brodkey, R. M. Christensen, S. Crandall, C. F. Dafermos, C. Dalton, D. C. Drucker, M. Fourney, D. Frederick, L. B. Freund, T. L. Geers, J. Goddard, J. W. Hutchinson, P. G. Hodge, N. J. Hoff, R. M. Jones, D. D. Joseph, A. Kobayashi, Y. H. Ku, L. G. Leal, S. Leibovich, J. T. Oden, P. Spanos, W. Vorus, F. Y. M. Wan, R. P. Wei

USNC/TAM Meeting: June 19 & 20, 1992, Washington, DC	
1993	J. D. Achenbach, A. Acrivos, H. Aref, T. Belytschko, M. M. Bernitsas, D. Bogy, B. A. Boley, R. S. Brodkey, R. M. Christensen, C. F. Dafermos, C. Dalton, E. H. Dowell, D. C. Drucker, M. Fourney, D. Frederick, L. B. Freund, T. L. Geers, P. G. Hodge, N. J. Hoff, G. Homsy, J. W. Hutchinson, R. M. Jones, D. D. Joseph, A. Kobayashi, A. Kraynik, Y. H. Ku, L. G. Leal, S. Leibovich, J. T. Oden, A. Pierce, P. Spanos, W. Vorus, F. Y. M. Wan, R. P. Wei

USNC/TAM Meeting: June 14, 1993, Washington, DC	
1994	J. D. Achenbach, A. Acrivos, H. Aref, T. Belytschko, M. M. Bernitsas, D. Bogy, B. A. Boley, R. S. Brodkey, R. M. Christensen, C. F. Dafermos, E. H. Dowell, D. C. Drucker, M. Fourney, D. Frederick, L. B. Freund, T. L. Geers, P. G. Hodge, N. J. Hoff, G. Homsy, J. W. Hutchinson, R. M. Jones, D. D. Joseph, A. Kraynik, S. Leibovich, A. Kobayashi, Y. H. Ku, L. G. Leal, J. Lumley, D. L. McDowell, J. T. Oden, A. D. Pierce, P. Spanos, F. Y. M. Wan

USNC/TAM Meeting: June 26, 1994, Seattle, Washington	
1995	J. D. Achenbach, A. Acrivos, R. J. Adrian, H. Aref, M. M. Bernitsas, D. B. Bogy, B. A. Boley, R. S. Brodkey, C. F. Dafermos, E. H. Dowell, D. C. Drucker, G. Dvorak, L. B. Freund, M. Fourney, T. L. Geers, P. G. Hodge, N. J. Hoff, J. W. Hutchinson, R. D. James, R. M. Jones, D. D. Joseph, A. M. Kraynik, A. Kobayashi, Y. H. Ku, L. G. Leal, J. Lumley, D. L. McDowell, J. T. Oden, A. D. Pierce, P. Spanos, F. Y. M. Wan

USNC/TAM Meeting: May 19 & 20, 1995, Washington, DC	
1996	J. D. Achenbach, A. Acrivos, R. J. Adrian, D. S. Ahluwalia, H. Aref, M. M. Bernitsas, D. B. Bogy, B. A. Boley, R. S. Brodkey, R. Brown, C. F. Dafermos, E. H. Dowell, D. C. Drucker, G. Dvorak, M. Eisenberg, M. Fourney, L. B. Freund, T. L. Geers, J. Goddard, J. C. Hill, P. G. Hodge, N. J. Hoff, J. W. Hutchinson, R. D. James, R. M. Jones, A. M. Kraynik, Y. H. Ku, L. G. Leal, D. L. McDowell, J. T. Oden, A. D. Pierce, P. Spanos

USNC/TAM Meeting: May 17 & 18, 1996, Washington, DC	
1997	J. D. Achenbach, R. J. Adrian, D. S. Ahluwalia, H. Aref, M. M. Bernitsas, D. B. Bogy, B. A. Boley, R. Brodkey, R. Brown, I. Busch-Vishniac, E. H. Dowell, D. C. Drucker, G. Dvorak, M. Eisenberg, M. Fourney, L. B. Freund, C. T. Herakovich, J. C. Hill, P. G. Hodge, Jr, P. Holmes, R. D. James, R. M. Jones, A. M. Kraynik, Y. H. Ku, L. G. Leal, D. L. McDowell, J. T. Oden, A. D. Pierce, P. Spanos

(continued)

(continued)

USNC/TAM Meeting: May 2 & 3, 1997, Washington, DC	
Note: Hoff died on August 4, 1997	
1998	J. D. Achenbach, R. J. Adrian, D. S. Ahluwalia, H. Aref, H. Armen, M. M. Bernitsas, D. B. Bogy, B. A. Boley, R. Brodkey, I. Busch-Vishniac, E. H. Dowell, D. C. Drucker, G. Dvorak, M. Eisenberg, M. Fourney, L. B. Freund, C. T. Herakovich, J. C. Hill, P. G. Hodge, Jr, P. J. Holmes, R. D. James, R. M. Jones, A. M. Kraynik, Y. H. Ku, L. G. Leal, D. L. McDowell, J. T. Oden, A. D. Pierce, A. Prosperetti, A. J. Smits, S. Sture
USNC/TAM Meeting: June 21, 1998, Gainesville, FL	
1999	J. D. Achenbach, R. J. Adrian, D. S. Ahluwalia, H. Aref, H. Armen, M. M. Bernitsas, D. B. Bogy, B. A. Boley, I. Busch-Vishniac, E. H. Dowell, D. C. Drucker, G. Dvorak, M. Eisenberg, W. L. Fourney, L. B. Freund, C. E. Harris, C. T. Herakovich, J. C. Hill, P. G. Hodge, Jr, P. J. Holmes, J. W. Hutchinson, A. M. Kraynik, S. Kyriakides, Y. H. Ku, L. G. Leal, D. L. McDowell, J. T. Oden, A. D. Pierce, A. Prosperetti, A. J. Smits, S. Sture
USNC/TAM Meeting: April 30 & May 1, 1999, Washington, DC	
2000	A. Acrivos, R. J. Adrian, H. Aref, M. M. Bernitsas, D. B. Bogy, B. A. Boley, E. H. Dowell, D. C. Drucker, G. Dvorak, W. L. Fourney, L. B. Freund, C. E. Harris, E. G. Henneke, C. T. Herakovich, J. C. Hill, P. G. Hodge, P. J. Holmes, J. W. Hutchinson, A. M. Kraynik, S. Kyriakides, L. G. Leal, D. L. McDowell, P. Moin, J. T. Oden, A. D. Pierce, A. Prosperetti, A. Smits, S. Sture, F. Y. M. Wan
USNC/TAM Meeting: August 27, 2000, Chicago, IL	
2001	R. J. Adrian, H. Aref, Z. Bažant, T. Belytschko, M. M. Bernitsas, D. B. Bogy, B. A. Boley, M. C. Boyce, R. Chona, E. H. Dowell, D. C. Drucker, G. Dvorak, W. L. Fourney, L. B. Freund, C. E. Harris, E. G. Henneke, C. T. Herakovich, J. C. Hill, P. G. Hodge, S. Kim, D. Kinderlehrer, W. G. Knauss, D. Krajcinovic, S. Kyriakides, Y. H. Ku, L. G. Leal, P. Moin, A. D. Pierce, A. Prosperetti, A. Smits, S. Sture, F. Y. M. Wan
USNC/TAM Meeting: April 27 & 28, 2001, Washington, DC	
Note: Drucker died on September 1, 2001	
2002	J. D. Achenbach, R. J. Adrian, H. Aref, N. Aubry, Z. Bažant, T. Belytschko, D. B. Bogy, B. A. Boley, M. C. Boyce, R. Chona, E. H. Dowell, G. Dvorak, W. L. Fourney, L. B. Freund, C. E. Harris, E. G. Henneke, C. T. Herakovich, J. C. Hill, P. G. Hodge, T. J. R. Hughes, D. G. Karr, S. Kim, D. Kinderlehrer, W. G. Knauss, D. Krajcinovic, S. Kyriakides, L. G. Leal, P. Moin, A. N. Norris, A. Smits, S. Sture, F. Y. M. Wan
Note: Ku died on September 9, 2002	
USNC/TAM Meeting: June 23, 2002, Blacksburg, VA	
2003	J. D. Achenbach, R. J. Adrian, H. Aref, N. Aubry, T. Belytschko, D. B. Bogy, B. A. Boley, R. Chona, L. B. Freund, C. E. Harris, E. G. Henneke, C. T. Herakovich, J. C. Hill, P. G. Hodge, T. J. R. Hughes, D. G. Karr, S. Kim, D. Kinderlehrer, W. G. W. Knauss, S. Kyriakides, L. G. Leal, M. E. Mear, P. Moin, A. N. Norris, O. O. Ochoa, K. Ravi-Chandar, A. Shukla, A. Smits, S. Sture, J. Wallace, F. Y. M. Wan

(continued)

(continued)

USNC/TAM Meeting: May 2 & 3, 2003, Washington, DC	
2004	R. Abeyaratne, J. D. Achenbach, A. Acrivos, R. J. Adrian, H. Aref, N. Aubry, L. C. Brinson (EO), T. Belytschko, D. B. Bogy, J. F. Brady, B. A. Boley, L. R. Collins, R. Chona, J. Fish, J. F. Foss, L. B. Freund, T. L. Geers, M. D. Graham, C. E. Harris, C. T. Herakovich, J. L. Higdon, P. G. Hodge, T. J. R. Hughes, W. D. Iwan, I. M. Jasiuk, D. G. Karr, S. Kim, D. Kinderlehrer, W. G. Knauss, S. Kyriakides, L. G. Leal, J. E. Marsden, M. E. Mear, A. N. Norris, O. O. Ochoa, K. Ravi-Chandar, A. Shukla, N. R. Sottos, S. Sture, Z. Suo, J. A. Todd, S. M. Troian, J. Wallace, F. Y. M. Wan

USNC/TAM Meeting: April 30 & May 1, 2004. Washington, DC	
2005	R. Abeyaratne, J. D. Achenbach, A. Acrivos, H. Aref, N. Aubry, L. C. Brinson (EO), T. Belytschko, J. F. Brady, B. A. Boley, L. R. Collins, R. Chona, J. Fish, J. F. Foss, L. B. Freund, T. L. Geers, M. D. Graham, C. E. Harris, C. T. Herakovich, J. L. Higdon, P. G. Hodge, T. J. R. Hughes, W. D. Iwan, I. M. Jasiuk, D. G. Karr, D. Kinderlehrer, W. G. Knauss, S. Kyriakides, L. G. Leal, J. E. Marsden, M. E. Mear, A. N. Norris, O. O. Ochoa, K. Ravi-Chandar, A. Shukla, N. R. Sottos, S. Sture, Z. Suo, J. A. Todd, S. M. Troian, J. Wallace, F. Y. M. Wan

USNC/TAM Meeting: April 29 & 30, 2005, Washington, DC	
2006	R. Abeyaratne, J. D. Achenbach, A. Acrivos, H. Aref, N. Aubry, L. C. Brinson (EO), T. Belytschko, J. F. Brady, B. A. Boley, L. R. Collins, R. Chona, J. Fish, J. F. Foss, L. B. Freund, T. L. Geers, M. D. Graham, C. E. Harris, C. T. Herakovich, J. L. Higdon, P. G. Hodge, T. J. R. Hughes, W. D. Iwan, I. M. Jasiuk, D. G. Karr, D. Kinderlehrer, W. G. Knauss, S. Kyriakides, L. G. Leal, M. E. Mear, A. N. Norris, O. O. Ochoa, K. Ravi-Chandar, A. Shukla, N. R. Sottos, S. Sture, Z. Suo, J. A. Todd, S. M. Troian, J. Wallace, F. Y. M. Wan

USNC/TAM Meeting: June 25, 2006, Boulder, CO	
2007	R. Abeyaratne, J. D. Achenbach, A. Acrivos, H. Aref, N. Aubry, L. C. Brinson (EO), T. Belytschko, J. F. Brady, B. A. Boley, L. R. Collins, W. W. Chen, R. Chona, J. Fish, J. F. Foss, L. B. Freund, T. L. Geers, M. D. Graham, C. E. Harris, C. T. Herakovich, J. L. Higdon, P. G. Hodge, T. J. R. Hughes, W. D. Iwan, I. M. Jasiuk, D. G. Karr, D. Kinderlehrer, W. G. Knauss, S. Kyriakides, L. G. Leal, A. N. Norris, O. O. Ochoa, K. Ravi-Chandar, A. Shukla, N. R. Sottos, S. Sture, Z. Suo, J. A. Todd, S. M. Troian, J. Wallace, F. Y. M. Wan

USNC/TAM Meeting: April 27 & 28, 2007, Washington, DC	
2008	J. D. Achenbach, A. Acrivos, H. Aref, N. Aubry, T. Belytschko, B. A. Boley, J. F. Brady, L. C. Brinson (EO), W. W. Chen, R. Chona, L. R. Collins, J. Fish, J. F. Foss, L. B. Freund, M. D. Graham, T. J. Healey, C. T. Herakovich, J. L. Higdon, P. G. Hodge, T. J. R. Hughes, W. D. Iwan, I. M. Jasiuk, D. G. Karr, D. Kinderlehrer, S. Kyriakides, W. K. Liu, G. H. McKinley, R. M. McKeeking, A. N. Norris, K. Ravi-Chandar, A. Shukla, Z. Suo, J. A. Todd

USNC/TAM Meeting: April 25 & 26, 2008, Washington, DC	
2009	J. D. Achenbach, A. Acrivos, H. Aref, N. Aubry, B. A. Boley, J. F. Brady, L. C. Brinson (EO), R. E. Caflisch, W. W. Chen, R. Chona, L. R. Collins, J. Fish, J. F. Foss, L. B. Freund, H. Gao, T. J. Healy, C. T. Herakovich, J. L. Higdon, T. J. R. Hughes, W. D. Iwan, I. M. Jasiuk, D. G. Karr, S. Kyriakides, L. G. Leal, W. K. Liu, G. H. McKinley, R. M. McMeeking, A. N. Norris, K. Ravi-Chandar, E. S. G. Shaqfeh, A. Shukla, K. J. Stebe, Z. Suo, J. A. Todd

(continued)

(continued)

USNC/TAM Meeting: April 24 & 25, 2009, Washington, DC
2010	J. D. Achenbach, A. Acrivos, H. Aref, N. Aubry, B. A. Boley, J. F. Brady, L. C. Brinson (EO), R. E. Caflisch, W. W. Chen, L. R. Collins, R. H. Dodds, J. Fish, J. F. Foss, L. P. Franzoni, L. B. Freund, H. Gao, T. J. Healy, C. T. Herakovich, J. L. Higdon, T. J. R. Hughes, W. D. Iwan, I. M. Jasiuk, S. Kyriakides, L. G. Leal, W. K. Liu, G. H. McKinley, R. M. McMeeking, K. Ravi-Chandar, E. S. G. Shaqfeh, A. Shukla, K. J. Stebe, Z. Suo, J. A. Todd, Y. L. J. Young

USNC/TAM Meeting: June 27, 2010, State College, PA
2011	J. D. Achenbach, A. Acrivos, H. Aref, A. Aubry, B. A. Boley, J. F. Brady, L. C. Brinson (EO), R. E. Caflisch, W. W. Chen, L. R. Collins, R. H. Dodds, H. D. Espinosa, J. Fish, J. F. Foss, L. P. Franzoni, L. B. Freund, H. Gao, T. J. Healey, C. T. Herakovich, J. L. Higdon, T. J. R. Hughes, W. D. Iwan, S. Kyriakides, L. G. Leal, W. K. Liu, G. H. McKinley, R. M. McMeeking, G. Ravichandran, K. Ravi-Chandar, E. S. G. Shaqfeh, H. Stone, K. J. Stebe, Z. Suo, J. A. Todd, Y. L. J. Young

USNC/TAM Meeting: May 6 & 7, 2011, Washington, DC
2012	J. D. Achenbach, A. Acrivos, A. Aubry, B. A. Boley, R. Caflisch, W. W. Chen, L. R. Collins, R. H. Dodds, H. D. Espinosa, J. Fish, J. F. Foss, L. P. Franzoni, L. B. Freund, H. Gao, T. J. Healey, C. T. Herakovich, T. J. R. Hughes, W. D. Iwan, A. R. Karagozian, S. Kyriakides, C. M. Landis, L. G. Leal, W. K. Liu, E. Longmire, R. M. McMeeking, G. H. McKinley, J. Morris, S. Peercy (EO) G. Ravichandran, K. Ravi-Chandar, E. S. G. Shaqfeh, K. J. Stebe, H. Stone, Z. Suo, Y. L. J. Youn

USNC/TAM Meeting: May 4 & 5, 2012, Washington, DC

Appendix E.2: USNC/TAM Officers

Term	Chair	Vice-Chair	Secretary	Past chair
1948–1956	H. L. Dryden			
1956–1958	N. J. Hoff	M. Hetenyi	C. E. Davies	
1958–1960	N. J. Hoff	W. Prager	O. B. Schier II	
1960–1962	W. Prager	M. Hetenyi	O. B. Schier II	
1962–1964	M. Hetenyi	F. N. Frenkiel	O. B. Schier II	
1964–1966	F. N. Frenkiel	D. C. Drucker	O. B. Schier II	
1966–1968	D. C. Drucker	G. F. Carrier	O. B. Schier II	
1968–1970	G. F. Carrier	E. H. Lee	O. B. Schier II	
1970–1972	E. H. Lee	G. F. Carrier	F. N. Frenkiel	
1972–1974	S. H. Crandall	E. H. Lee	F. N. Frenkiel	
1974–1976	B. A. Boley	S. H. Crandall	F. N. Frenkiel	
1976–1978	R. S. Rivlin	B. A. Boley	F. N. Frenkiel	
1978–1980	P. M. Naghdi	R. S. Rivlin	F. N. Frenkiel	
1980–1982	J. B. Keller	P. M. Naghdi	F. N. Frenkiel	
1982–1984	J. W. Dally	J. B. Keller	P. G. Hodge, Jr.	
1984–1986	A. Acrivos	J. W. Dally	P. G. Hodge, Jr.	
1986–1988	H. N. Abramson	A. Acrivos	P. G. Hodge, Jr.	
1988–1990	R. M. Christensen	H. N. Abramson	P. G. Hodge, Jr.	

(continued)

(continued)

Term	Chair	Vice-Chair	Secretary	Past chair
1990–1992	S. Leibovich	R. M. Christensen	P. G. Hodge, Jr.	
1992–1994	J. T. Oden	S. Leibovich	P. G. Hodge, Jr.	
1994–1995	L. G. Leal	J. T. Oden	P. G. Hodge, Jr.	
1995–1996	L. G. Leal	R. Brodkey	P. G. Hodge, Jr.	
1996–1998	E. H. Dowell	R. Brodkey	P. G. Hodge, Jr.	L. G. Leal
1998–2000	R. J. Adrian	J. C. Hill	P. G. Hodge, Jr.	E. H. Dowell
2000–2002	H. Aref	J. C. Hill	C. T. Herakovich	R. J. Adrian
2002–2004	W. G. Knauss	D. B. Bogy	C. T. Herakovich	H. Aref
2004–2006	T. Belytschko	N. Aubry	C. T. Herakovich	W. G. Knauss
2006–2008	N. Aubry	T. J. R. Hughes	C. T. Herakovich	T. Belytschko
2008–2010	T. J. R. Hughes	L. Collins	C. T. Herakovich	N. Aubry
2010–2012	L. Collins	S. Kyriakides	C. T. Herakovich	T. J. R. Hughes
2012–2014	S. Kyriakides	W. K. Liu	L. P. Franzoni	L. Collins

Appendix E.3: USNC/TAM Society Representatives

The superscripts in Appendices E.3 and E.4 refer to the footnotes listed after Appendix E.3.

Term	AIChE	APS	SIAM	
1959–1963	A. Acrivos	F. N. Frenkiel		
1963–1967	A. Acrivos	F. N. Frenkiel		
1967–1971	A. Acrivos	G. B. Schubauer		
1971–1975	W. R. Schowalter	F. N. Frenkiel	G. H. Handelman[1]	
1975–1979	W. R. Schowalter	F. N. Frenkiel	G. H. Handelman	
1979–1983	T. J. Hanratty	R. L. Peskin	S. H. Davis	
1983–1987	T. J. Hanratty	S. H. Davis	R. C. DiPrima	
1987–1991	R. S. Brodkey	M. D. Van Dyke	M. Tulin[6]	
1991–1995	R. S. Brodkey	L. G. Leal[8]	F. Y. M. Wan	
1995–1999	J. C. Hill	H. Aref	D. S. Ahluwalia	
1999–2003	J. C. Hill	P. Moin[16]	D. S. Ahluwalia/ F. Y-M. Wan[17]	
2003–2007	J. Higdon	J. F. Foss[21]	F. Y. M. Wan	
2007–2011	J. Higdon	J. F. Foss	T. J. Healey	
2011–2015	J. Morris	E. Longmire	T. J. Healey	
Term	AIAA	SEM[13]	SES	USACM
1958–1962	N. J. Hoff	W. Ramberg		
1962–1966	N. J. Hoff	D. C. Drucker		
1966–1970	N. J. Hoff	D. C. Drucker		
1970–1974	B. Budiansky	D. C. Drucker		
1974–1978	B. Budiansky	J. W. Dally[2]		
1978–1982	H. N. Abramson	J. W. Dally		
1982–1986	H. N. Abramson	C. E. Taylor	R. T. Shield[7]	
1986–1990	B. Noton	C. E. Taylor	D. Frederick	

(continued)

(continued)

Term			
1990–1994	R. M. Jones[9]	M. Fourney	D. Frederick	
1994–1998	R. M. Jones	M. Fourney	G. Dvorak	
1998–2002	C. Harris	W. Fourney	G. Dvorak	
2002–2006	C. Harris	A. Shukla	M. E. Mear	
2006–2010	W. Chen	A. Shukla	I. Jasiuk	J. Fish
2010–2014	W. Chen	G. Ravichandran	H. Espinosa	J. Fish

Term	ASCE	ASTM	SNAME	ASA
1961–1965	D. C. Drucker	T. J. Dolan		
1965–1969	E. P. Popov	T. J. Dolan		
1969–1973	E. P. Popov	T. J. Dolan		
1973–1977	M. L. Baron	P. C. Paris[3]		
1977–1981	M. L. Baron	P. C. Paris		
1981–1985	J. L. Sackman	J. W. Hutchinson	R. D. Cooper[4]	
1985–1989	T. Belytschko	J. W. Hutchinson	R. D. Cooper	
1989–1993	P. Spanos	R. Wei	W. Vorus	
1993–1997	P. Spanos	D. McDowell[13]	M. Bernitsas[10]	A. Pierce[12]
1997–2001	S. Sture	D. McDowell	M. Bernitsas	A. Pierce
2001–2005	S. Sture	R. Chona	D. G. Karr	A. Norris
2005–2009	W. D. Iwan	R. Chona	D. G. Karr	A. Norris
2009–2013	W. D. Iwan	R. H. Dodds Jr.	Y. L. Young	L. P. Franzoni

Term	ASME	AMS	SOR	AAM
1960–1964	A. M. Wahl	W. Prager	E. H. Lee	
1964–1968	G. F. Carrier	W. Prager	E. H. Lee	
1968–1972	S. H. Crandall	J. B. Keller	E. H. Lee	
1972–1976	P. M. Naghdi	J. B. Keller	R. S. Rivlin	
1976–1980	J. Miklowitz	L. N. Howard	H. Markovitz	
1980–1984	R. Skalak	S. S. Antman	H. Markovitz	
1984–1988	R. M. Christensen	S. S. Antman	J. Goddard	
1988–1992	T. L. Geers	C. M. Dafermos	J. Goddard	
1992–1996	T. L. Geers	C. M. Dafermos	A. M. Kraynik	E. H. Dowell[11]
1996–2000	C. T. Herakovich	P. Holmes	A. M. Kraynik	E. H. Dowell
2000–2004	D. Krajcinovic/ S. Kyriakides[19] D. Kinderlehrer	S. Kim	Z. Bazant/	
K. RaviChandar[20]				
2004–2008	S. Kyriakides	J. E. Marsden/ D. Kinderlehrer[22]	M. D. Graham	K. Ravi-Chandar
2008–2012	S. Kyriakides	R. Caflisch	E. Shaqfeh	K. Ravi-Chandar
2012–2016	N. Aubry	E. Tadmor	E. Shaqfeh	C. Landis

Footnotes

The footnotes are for the Society Representatives listed above and the Members-at-Large that follow

1. G. H. Handelman appointed as representative from SIAM in February 1972
2. J. W. Dally appointed in December 1973 to complete D. C. Drucker's term

3. P. C. Paris appointed in March 1973 to complete T. J. Dolan's term
4. R. D. Cooper appointed as representative from SNAME in April 1982
5. W. E. Schowalter appointed November 1979 to complete S. H. Davis' term
6. M. Tulin appointed in January 1985 to complete R. C. DiPrima's term
7. R. T. Shield appointed as representative from SES in November 1985
8. L. G. Leal appointed in November 1990 to complete M. Van Dyke's term
9. R. M. Jones appointed in April 1992 to complete B. Noton's term
10. M. Bernitsas appointed in February 1993 to complete W. Vorus' term
11. E. Dowell appointed as representative from AAM in May 1993
12. A. Pierce appointed as representative from ASA in May 1993
13. D. McDowell appointed in May 1994 to complete R. Wei's second term
14. D. Bogy re-elected to complete C. Mei's term, May 1996
15. A. Smits elected to complete R. Brown's second term, June 1998, eligible for 2+ terms
16. P. Moin appointed in April 1999 to complete H. Aref's term
17. F. Y. M. Wan appointed on April 9, 2001 to replace Ahluwalia who resigned
18. T. J. R. Hughes elected on Apr 28, 2001 replacing J. Hutchinson who resigned. Eligible for 2+ terms
19. S. Kyriakides replaced Krajcinovic for 2003–2004. Appointed first term for 04-08
20. K. Ravi-Chandar replaced Z. Bazant on 2/25/03. Bazant resigned as AAM representative
21. Nominated to fill Moin seat on March 16, 2004. Moin not re-appointed
22. D. Kinderlehrer nominated to replace J. E. Marsden who resigned effective Oct. 31, 2005

Appendix E.4: USNC/TAM Members-at-Large

Other than for the year 1948, these data are available only for years 1969–2012. MAL terms changed from 2 to 3 years in 2008. Refer to section "USNC/TAM Society Representatives" for footnotes.

Odd/even year ending MAL terms			
1973–1975	J. D. Achenbach	J. R. Rice	J. D. Cole	D. J. Joseph
1975–1977	J. D. Achenbach	J. R. Rice	J. L. Sackman	P. G. Saffman
1977–1979	H. Brenner	E. Murman	A. Roshko	R. A. Toupin
1979–1981	R. M. Christensen	J. W. Miles	A. Roshko	R. A. Toupin
1981–1983	R. M. Christensen	J. W. Miles	S. Corrsin	R. T. Shield
1983–1985	E. B. Becker	P. A. Libby	S. Corrsin	R. T. Shield
1985–1987	J. P. Lamb	P. A. Libby	L. B. Freund	C. S. Hsu
1987–1989	J. P. Lamb	C. F. Chen	L. B. Freund	C. S. Hsu
1989–1991	C. Dalton	C. F. Chen	L. B. Freund	D. D. Joseph
1991–1993	C. Dalton	A. Kobayashi	L. B. Freund	D. D. Joseph
1993–1995	J. Lumley	A. Kobayashi	L. B. Freund	D. D. Joseph
1995–1997	R. Brown	M. Eisenberg	L. B. Freund	C. Mei/D. B. Bogy[14]
1997–1999	A. J. Smits[15]	M. Eisenberg	H. Armen	A. Prosperetti
1999–2001	A. J. Smits	E. G. Henneke	W. G. Knauss	A. Prosperetti
2001–2003	A. J. Smits	E. G. Henneke	W. G. Knauss	N. Aubry
2003–2005	N. R. Sottos	T. Geers	R. Abeyaratne	N. Aubry
2005–2007	N. R. Sottos	R. Abeyaratne	S. M. Troian	

(continued)

(continued)

2008–2011	J. F. Brady	K. J. Stebe	Z. Suo
2011–2014	C. M. Landis	K. J. Stebe	A. R. Karagozian
2013–2016	J. O. Dabiri		

Even/odd year ending MAL terms

1969–1970	H. Rouse	H. W. Liepmann	W. Prager
1970–1972	P. S. Klebanoff	H. W. Liepmann	W. Prager
1972–1974	P. S. Klebanoff	H. W. Liepmann	W. Prager
1974–1976	M. V. Morkovin	F. Essenburg	H. Brenner
1976–1978	E. H. Lee	F. Essenburg	P. M. Naghdi
1978–1980	B. Budiansky	J. B. Keller	S. H. Davis/W. E. Schowalter[5]
1980–1982	A. Acrivos	M. M. Carroll	Y. H. Pao
1982–1984	A. Acrivos	M. M. Carroll	Y. H. Pao
1984–1986	R. Nordgren	R. Plunkett	S. Widnall
1986–1988	J. T. Oden	R. Plunkett	S. Widnall
1988–1990	J. T. Oden	D. B. Bogy	S. Leibovich
1990–1992	J. T. Oden	D. B. Bogy	T. Belytschko
1992–1994	G. Homsy	D. B. Bogy	T. Belytschko
1994–1996	R. J. Adrian	D. B. Bogy	R. James
1996–1998	R. J. Adrian	I. Busch-Vishniac	R. James
1998–2000	J. W. Hutchinson	I. Busch-Vishniac	S. Kyriakides
2000–2002	J. W. Hutchinson/T. J. R. Hughes[18]	M. Boyce	S. Kyriakides
2002–2004	T. J. R. Hughes	O. O. Ochoa	J. Wallace
2004–2006	T. J. R. Hughes	O. O. Ochoa	J. Wallace
2006–2008	J. F. Brady	L. R. Collins	Z. Suo
2007–2010	W. K. Liu	G. McKinley	R. McMeeking
2010–2013	W. K. Liu	G. McKinley	R. McMeeking

Appendix E.5: U.S. Congress Chairs

Ex-official voting member as per change in constitution in November 2002

2003–2007	T. L. Geers (Co-Chair with S. Sture for 2006 U.S. Congress)
2007–2011	J. A. Todd (Co-Chair with C. Bakis for 2010 U.S. Congress)
2011–2015	J. F. Foss (Co-Chair with T. Pence for 2014 U.S. Congress)

Appendix F: USNC/TAM Resolutions

© Springer International Publishing Switzerland 2016
C.T. Herakovich, *Mechanics IUTAM USNC/TAM*,
DOI 10.1007/978-3-319-32312-1

Frenkiel Resolution

United States National Committee on Theoretical and Applied Mechanics
of the
National Academy of Sciences

he U.S. National Committee on Theoretical and Applied Mechani.
luring its regular meeting on April 21, 1983, voted unanimous
in favor of the following resolution:

BE IT RESOLVED, that

Dr. Francois N. Frenkiel

is herewith highly commended for his long standing anc
effective service to the Mechanics Community as a membe
chairman, and secretary of this Committee.

THEREFORE, the U.S. National Committee on Theoretical an.
Applied Mechanics extends to Dr. Frenkiel its most sincer
congratulations and best wishes for good health and a
active life following the occasion of his retirement.

With grateful appreciation,

Philip G. Hodge Jr.
Secretary

James W. Dally
Chairman

Hodge Resolution

Resolution

U.S. National Committee on Theoretical and Applied Mechanics

Whereas, Philip G. Hodge is Professor Emeritus of Mechanics at the University of Minnesota and Visiting Professor Emeritus, Division of Applied Mechanics, Stanford University, and also served at Brown University, University of California at Los Angeles, Polytechnic Institute of Brooklyn, and Illinois Institute of Technology, and

Whereas, Philip G. Hodge has authored or co-authored five books, published numerous archival papers in mechanics, and served as Editor of the ASME Journal of Applied Mechanics for 5 years, and

Whereas, Philip G. Hodge's achievements have been recognized with his election to the National Academy of Engineering, and honored with the ASME Medal, the ASME Worcester Reed Warner Medal, the ASME Daniel C. Drucker Medal, the ASCE Theodore van Kármán Medal, the USSR Euler Medal, and Honorary Membership in ASME, and

Whereas, Philip G. Hodge has given outstanding and dedicated service to the U.S. National Committee on Theoretical and Applied Mechanics since 1982, including 18 years as Secretary, and to the International Union of Theoretical and Applied Mechanics since 1984,

Now, Therefore, Be It Resolved that the U.S. National Committee on Theoretical and Applied Mechanics, on behalf of its members and the entire mechanics community, on this day, April 25, 2008, express their appreciation and admiration to Philip G. Hodge for his service and extend to him their very best wishes.

Signed:

Nadine Aubry, Chair Carl T. Herakovich, Secretary
Herakovich Plaque

Appendix G: USNC/TAM Members in Elected IUTAM Positions

U.S. Bureau Members and Members-at-Large

H. L. Dryden	1948–1965
N. J. Hoff	1972–1997
F. N. Frenkiel	1980–1988
D. C. Drucker	1988–2001
B. A. Boley	1980–2016
P. G. Hodge, Jr.	1996–2008
A. Acrivos	2004–2016
J. D. Achenbach	2009–2016
L. B. Freund	2012–2016
J. M. Burgers	1948–1981
Y. H. Ku	1948–2002
T. von Kármán	1948–1963
S. P. Timoshenko	1948–1972
J. D. Hunsaker	1948–1984
R. von Mises	1948–1953
S. Goldstein	1948–1989

U.S. Members of the IUTAM Bureau

IUTAM has a policy that only one person from any country may be on the eight-member IUTAM Bureau at any time. The Bureau consists of four officers (President, Vice-President, Treasurer & Secretary General) and four Members. The United States has had a member of the Bureau for every year since establishment of IUTAM in 1948. These people and their positions are listed below. J. M. Burgers was not from the United States when he was IUTAM Secretary from 1948 to 1952 and a Member from 1952 to 1956.

© Springer International Publishing Switzerland 2016
C.T. Herakovich, *Mechanics IUTAM USNC/TAM*,
DOI 10.1007/978-3-319-32312-1

1948–1960—Hugh Dryden: Treasurer (1948–1952), President (1952–1956), Vice-President (1956–1960).

1960–1972—Nick Hoff: Member.

1972–1988—Dan Drucker: Treasurer (1972–1980), President (1980–1984), Vice-President (1984–1988).

1988–1996—Bruno Boley: Member (1988–1994), Treasurer (1994–1996).

1996–2012—Ben Freund:Treasurer (1996–2004), President (2004–2008), Vice-President (2008–2012).

2012–2016—Nadine Aubry:Member.

U.S. Members of the IUTAM Congress Committee
Records are available only beginning in 1965.

J. D. Achenbach	1986–1994
A. Acrivos	1988–2000
H. Aref	1994–2011
N. Aubry	2004–2012
T. Belytschko	2002–2006
D. B. Bogy	1996–2002
B. A. Boley	1968–1996
J. M. Burgers	1965–1980
J. D. Cole	1974–1980
D. C. Drucker	1980–1992
H. C. Dryden	1965–1967
H. D. Espinosa	2012–2016
L. B. Freund	1996–2008
H. Gao	2012–2016
S. Goldstein	1965–1975
J. P. den Hartog	1965–1967
C. T. Herakovich	2006–2014
N. J. Hoff	1968–1996
J. W. Hutchinson	1992–1996
A. R. Karagozian	2012–2016
Y. H. Ku	1976–2002
S. Kyriakides	2004–2012
L. G. Leal	2000–2012
S. Leibovich	1988–1996
R. M. McMeeking	2008–2016
J. T. Oden	1996–2002
K. Ravi-Chandar	2006–2014
E. S. G. Shaqfeh	2012–2016
H. Stone	2010–2014
S. P. Timoshenko	1965–1971

U.S. Members of the IUTAM Congress Executive Committee

1978–1988	B. Boley, Secretary (1978–1984)
1988–2000	A. Acrivos
2000–2011	H. Aref, Secretary (2008–2011) (died in office)
2012–2016	R. McMeeking, Secretary (2012–2016)

U.S. Members of the IUTAM Fluids Symposium Panel

1979–1984	F. N. Frenkiel
1984–2000	A. Acrivos, Chair (1988–2000)
2000–2016	L. G. Leal, Chair (2008–2016)

U.S. Members of the IUTAM Solids Symposium Panel

1978–1992	S. H. Crandall, Chair (1984–1992)
1988–1908	J. D. Achenbach, Chair (2004–12)
2008–2016	H. Gao

Appendix H: U.S. Members of the IUTAM General Assembly

These data are from the IUTAM Annual Reports. P denotes President, HP denotes Honorary President, T denotes Treasurer, V denotes Vice-President, and B denotes Board Member.

1948	H. L. Dryden-T, J. C. Hunsaker, Th. von Kármán, S. P. Timoshenko, R. von Mises
1949	H. L. Dryden-T, J. C. Hunsaker, Th. von Kármán, S. P. Timoshenko, R. von Mises, H. W. Emmons, N. J. Hoff, R. D. Mindlin, M. G. Salvadori, R. J. Seeger
1950	H. L. Dryden-T, J. C. Hunsaker, Th. von Kármán, S. P. Timoshenko, R. von Mises, H. W. Emmons, N. J. Hoff, R. D. Mindlin, M. G. Salvadori, R. J. Seeger
1951	H. L. Dryden-T, J. C. Hunsaker, Th. von Kármán, Y. M. Ku, S. P. Timoshenko, R. von Mises, H. W. Emmons, N. J. Hoff, R. D. Mindlin, M. G. Salvadori, R. J. Seeger
Note: Th. von Kármán was elected Honorary President of IUTAM effective from May 11, 1951 in recognition of his "outstanding service to the science of Mechanics and his great interest and help given in matters concerning the Union"	
1952	H. L. Dryden-T, J. C. Hunsaker, Th. von Kármán-HP, Y. M. Ku, S. P. Timoshenko, R. von Mises, H. W. Emmons, N. J. Hoff, A. T. Ippen, M. H. Martin, R. D. Mindlin, N. M. Newmark
1953	H. L. Dryden-P, J. C. Hunsaker, Th. von Kármán-HP, S. Timoshenko, R. von Mises, Y. M. Ku, N. J. Hoff, A. T. Ippen, M. H. Martin, R. D. Mindlin, N. M. Newmark
R. von Mises died in 1953	
1954–1956	H. L. Dryden-P, J. M. Burgers, F. N. Frenkiel, S. Goldstein, N. J. Hoff, A. T. Ippen, Th. von Kármán-HP, R. D. Mindlin, N. M. Newmark S. Timoshenko, Y. M. Ku
1957–1958	H. L. Dryden-V, J. M. Burgers, S. Goldstein, N. J. Hoff, M. Hetenyi, Th. von Kármán-HP, Y. M. Ku, S. Timoshenko, W. Prager, W. Ramberg, F. N. Frenkiel
1959–1960	J. M. Burgers, H. L. Dryden-V, F. N. Frenkiel, S. Goldstein, M. Hetenyi, N. J. Hoff, Th. von Kármán-HP, S. Timoshenko, Y. H. Ku, W. Prager, W. Ramberg
1961–1962	J. M. Burgers, H. L. Dryden, S. Goldstein, Th. von Kármán-HP, S. Timoshenko, Y. H. Ku, N. J. Hoff-B, W. Prager, M. Hetenyi, F. N. Frenkiel, W. Ramberg

(continued)

© Springer International Publishing Switzerland 2016
C.T. Herakovich, *Mechanics IUTAM USNC/TAM*,
DOI 10.1007/978-3-319-32312-1

(continued)

1963	N. J. Hoff-B, M. Hetenyi, F. N. Frenkiel, W. Prager, W. Ramberg, J. M. Burgers, H. L. Dryden, S. Goldstein, Th. von Kármán-HP, S. Timoshenko, Y. H. Ku
von Kármán died on May 6, 1963	
1964–1965	N. J. Hoff-B, D. C. Drucker, F. N. Frenkiel, W. Prager, G. F. Carrier, J. M. Burgers, H. L. Dryden, S. Goldstein, S. Timoshenko, Y. H. Ku
Dryden died on Dec. 2, 1965	
1966–1967	D. C. Drucker, G. F. Carrier, F. N. Frenkiel, N. J. Hoff-B, Y. H. Ku, W. Prager, S. Goldstein, S. Timoshenko, J. M. Burgers
1968	D. C. Drucker, G. F. Carrier, N. J. Hoff-B, S. Goldstein, W. Prager, S. Timoshenko, J. M. Burgers, Y. H. Ku, F. N. Frenkiel
1969	D. C. Drucker, G. F. Carrier, N. J. Hoff-B, S. Goldstein, E. H. Lee, W. Prager, S. Timoshenko, J. M. Burgers, Y. H. Ku, F. N. Frenkiel
1970–1971	N. J. Hoff-B, D. C. Drucker, E. H. Lee, F. N. Frenkiel, S. Goldstein, Y. H. Ku, W. Prager, S. Timoshenko, J. M. Burgers
Timoshenko died on May 29, 1972	
1972–1973	D. C. Drucker-T, S. H. Crandall, E. H. Lee, B. Budiansky, S. Goldstein, Y. H. Ku, F. N. Frenkiel, J. M. Burgers, N. J. Hoff
1974–1975	D. C. Drucker-T, J. M. Burgers, S. H. Crandall, E. H. Lee, N. J. Hoff, R. Rivlin, B. Budiansky, F. N. Frenkiel, S. Goldstein, Y. H. Ku
1976–1977	D. C. Drucker-T, R. S. Rivlin, B. A. Boley, S. Crandall, N. J. Hoff, J. M. Burgers, S. Goldstein, Y. H. Ku, F. N. Frenkiel
1978–1979	D. C. Drucker-T, P. M. Naghdi, R. S. Rivlin, B. A. Boley, N. J. Hoff, J. M. Burgers, S. Goldstein, Y. H. Ku, F. N. Frenkiel
1980–1981	D. C. Drucker-P, B. A. Boley, J. B. Keller, P. M. Naghdi, R. S. Rivlin, N. J. Hoff, J. M. Burgers, S. Goldstein, Y. H. Ku, F. N. Frenkiel
Burgers died on June 7, 1981	
1982–1983	B. A. Boley, D. C. Drucker-P, J. W. Dally, R. C. DiPrima, J. B. Keller, P. G. Hodge, Jr., N. J. Hoff, S. Goldstein, Y. H. Ku, P. M. Naghdi, F. N. Frenkiel, R. S. Rivlin
1984	B. A. Boley, J. W. Dally, R. C. DiPrima, D. C. Drucker-V, F. N. Frenkiel, S. Goldstein, P. G. Hodge, Jr., N. J. Hoff, J. B. Keller, Y. H. Ku, P. M. Naghdi
Hunsaker died in September 1984	
1985–1986	A. Acrivos, B. Boley, J. W. Dally, S. H. Davis, R. C. DiPrima, D. C. Drucker-V, F. N. Frenkiel, S. Goldstein, N. J. Hoff, P. G. Hodge, Jr., Y. H. Ku
1987	IUTAM Annual Report not available
1988–1990	H. N. Abramson, A. Acrivos, B A Boley, R. M. Christensen, D. C. Drucker, L. B. Freund, S. Goldstein, P. G. Hodge, Jr., N. J. Hoff, Y. H. Ku
Goldstein died on 22 January, 1989	
1991	B. A. Boley-B, R. M. Christensen, D. C. Drucker, L. B. Freund, P. G. Hodge, Jr., N. J. Hoff, D. Joseph, Y. H. Ku, L. G. Leal, S. Leibovich
1992–1993	B. A. Boley-B, D. C. Drucker, L. B. Freund, P. G. Hodge, Jr., N. J. Hoff, Y. H. Ku, L. G. Leal, S. Leibovich
1994 1995	H. Aref, B. A. Boley-T, D. C. Drucker, L. B. Freund, P. G. Hodge, Jr., N. J. Hoff, Y. H. Ku, L. G. Leal
1996–1997	J. Achenbach, H. Aref, B. A. Boley, E. H. Dowell, D. C. Drucker, L. B. Freund-T, P. G. Hodge, Jr., N. J. Hoff, Y. H. Ku, L. G. Leal

(continued)

(continued)

Hoff died on August 4, 1997	
1998–1999	R. J. Adrian, H. Aref, B. A. Boley, S. H. Crandall, E. H. Dowell, D. C. Drucker, L. B. Freund-T, C. T. Herakovich, P. G. Hodge, Jr., Y. H. Ku
2000	R. J. Adrian, H. Aref, T. Belytschko, B. A. Boley, E. H. Dowell, D. C. Drucker, L. B. Freund-T, C. T. Herakovich, P. G. Hodge, Jr, Y. H. Ku, L. G. Leal
2001	R. J. Adrian, H. Aref, T. Belytschko B. A. Boley, D. C. Drucker, L. B. Freund-T, C. T. Herakovich, P. G. Hodge, Jr, Y. H. Ku, L. G. Leal
Drucker died on Sept. 1, 2001	
Ku died on Sept. 9, 2002	
2002–2003	H. Aref, T. Belytschko, B. A. Boley, L. B. Freund-T, C. T. Herakovich, P. G. Hodge, Jr, W. G. Knauss, L. G. Leal
2004–2006	A. Acrivos, H. Aref, T. Belytschko, B. A. Boley, L. B. Freund-P, C. T. Herakovich, P. G. Hodge, Jr., W. G. Knauss, L. G. Leal
Note: U.S. Representative terms were altered to begin on odd years beginning in 2006	
2007–2008	A. Acrivos, H. Aref, N. Aubry, B. A. Boley, L. B. Freund-P, C. T. Herakovich, P. G. Hodge, Jr, L. G. Leal, Z. Suo
2009–2010	J. Achenbach, A. Acrivos, N. Aubry, B. A. Boley, L. R. Collins, L. B. Freund-V, C. T. Herakovich, S. Kyriakides, Z. Suo
2010–2011	J. Achenbach, A. Acrivos, N. Aubry, B. A. Boley, L. Franzoni, L. B. Freund-V, C. T. Herakovich, S. Kyriakides, Z. Suo
Note: Hassan Aref died on September 9, 2011	
2012–2013	J. Achenbach, A. Acrivos, N. Aubry-B, B. A. Boley, L. Franzoni, L. B. Freund, H. Gao, C. T. Herakovich, S. Kyriakides, K. Ravi-Chandar

Appendix I: Academic Genealogy

Last name	First name	Ph.D. University	Advisor
Abeyaratne	Rohan	Caltech	James K. Knowles
Abramson	H. Norman	Texas	Harold J. Plass
Achenbach	Jan	Stanford	C. C. Chao
Acrivos	Andreas	Minnesota	Neal Amundson
Adrian	Ronald	Cambridge	Allen Townsend
Ahluwalia	Daljit	Indiana	T. Y. Thomas
Antman	Stuart	Minnesota	William H. Warner
Aref	Hassan	Cornell	E. D. Siggia
Armen	Harry	NYU	Edward Wilson
Aubry	Nadine	Cornell	John Lumley
Babiri	John	Caltech	Morteza Gharib
Baron	Melvin	Columbia	Hans Bleich
Bažant	Zdeněk	Czech. Acad. Sci.	No advisor
Becker	Eric	UC Berkeley	Jerry Sackman
Belytschko	Ted	IIT	Philip G. Hodge
Bernitsas	Michael	MIT	C. Chryssostomidis
Bing	R.	Texas	Robert L. Moore
Bogy	David	Brown	Eli Sternberg
Boley	Bruno	Brooklyn Poly	Nicolas Hoff
Boyce	Mary	MIT	D. Parks & Ali Argon
Brady	John	Stanford	Andreas Acrivos
Brenner	Howard	NYU (ScD)	John Happel
Brinson	L. Catherine	Caltech	Wolfgang G. Knauss
Brodkey	Robert	Wisconsin	William R. Marshall
Brown	Robert	Minnesota	L. E. Scriven
Budiansky	Bernard	Brown	William Prager
Burgers	Jan	Leiden	Paul Ehrenfest
Busch-Vishniac	Ilene	MIT	Richard Lyon

(continued)

© Springer International Publishing Switzerland 2016
C.T. Herakovich, *Mechanics IUTAM USNC/TAM*,
DOI 10.1007/978-3-319-32312-1

(continued)

Last name	First name	Ph.D. University	Advisor
Caflisch	Russel	UCLA	George Papanicolaou
Carrier	George	Cornell	Norman Goodier
Carroll	Michael	Brown	Ronald S. Rivlin
Chen	Chuan (Tony)	Brown	Joseph H. Clarke
Chen	Weinong	Caltech	G. Ravichandran
Chona	Ravinder	Maryland	George R. Irwin
Christensen	Richard	Yale	Dana Young
Cole	Julian	Caltech	Paco Lagerstrom
Collins	Lance	Pennsylvinia	Stuart Churchill
Cooper	Ralph	Illinois	H. H. Korst
Corrsin	Stanley	Caltech	Th. von Kármán
Crandall	Stephen	MIT	J. P. den Hartog
Dafermos	Constantine	Johns Hopkins	J. L. Erickson
Dally	James	IIT	August Durelli
Dalton	Charles	Texas	Frank Masch
Davis	Stephen	RPI	Lee A. Segel
den Hartog	Jacob	Pittsburgh	Ludwig Prandtl
DiPrima	Richard	Carnegie Mellon	G. Handelman
Dodds Jr.	Robert	Illinois	Leonard A. Lopez
Dolan	Thomas	Illinois	no Ph.D.
Dowell	Earl	MIT	John Dugundji
Drew	Thomas	MIT	no Ph.D.
Drucker	Daniel	Columbia	Raymond D. Mindlin
Dryden	Hugh	Johns Hopkins	Joseph Ames
Dvorak	George	Brown	Daniel C. Drucker
Eisenberg	Martin	Yale	Aris Phillips
Emmons	Howard	Harvard	John Smith & Charles Berry
Espinosa	Horacio	Brown	R. Clifton & M. Ortiz
Essenburg	Frank	Michigan	Paul M. Naghdi
Fish	Jacob	Northwestern	Ted Belytschko
Foss	John	Purdue	J. B. Jones
Fourney	William	Illinois	Morris Stern
Fourney	Michael	Caltech	Albert Ellis
Franzoni	Linda	Duke	Earl H. Dowell
Frederick	Daniel	Michigan	Paul M. Naghdi
Frenkiel	Francois	Lille	J. Kampe de Feriet
Freund	L. Ben	Northwestern	Jan D. Achenbach
Gao	Huajian	Harvard	James R. Rice
Geers	Thomas	MIT	J. P. den Hartog
Goddard	Joe	UC Berkeley	Andreas Acrivos
Goldstein	Sidney	Cambridge	Harold Jeffreys
Graham	Michael	Cornell	Paul H. Steen

(continued)

(continued)

Last name	First name	Ph.D. University	Advisor
Handelman,	G.	Brown	William Prager
Hanratty	Thomas	Princeton	Richard Wilhelm
Harris	Charles	Virginia Tech	Don H. Morris
Healey	Timothy	Illinois	Robert Muncaster
Henneke	Edmund	Johns Hopkins	Robert E. Green
Herakovich	Carl	IIT	Philip G. Hodge
Hetenyi	Miklos	Michigan	Stephen P. Timoshenko
Higdon	Jonathan	Cambridge	James Lighthill
Hill	James	Washington	Charles Sleicher
Hodge	Philip	Brown	William Prager
Hoff	Nicholas	Stanford	Stephen P. Timoshenko
Holmes	Philip	Southampton (UK)	Robert G. White
Homsy	George (Bud)	Illinois	John Hudson
Howard	Louis	Princeton	Donald C. Spencer
Hsu	Chieh-Su	Stanford	Norman Goodier
Hughes	Thomas	U. C. Berkeley	Jacob Lubliner
Hunsaker	Jerome	MIT	no advisor
Hutchinson	John	Harvard	Bernie Budiansky
Ippen	Arthur	Caltech	Th von Kármán & R. T. Knapp
Iwan	Wilfred (Bill)	Caltech	Thomas K. Caughey
James	Richard	Johns Hopkins	Jerald L. Ericksen
Jasiuk	Iwona	Northwestern	Toshio Mura
Jones	Robert	Illinois	Arthur P. Boresi
Joseph	Daniel	IIT	L. N. Tao
Karagozian	Ann	Caltech	Frank E. Marble
Karr	Dale	Tulane	Sandar C. Das
Keller	Joseph	NYU	Richard Courant
Kim	Sangtae	Princeton	William B. Russel
Kinderlehrer	David	U C Berkeley	Hans Lewy
Klebanoff	P.	no Ph.D.	no advisor
Knauss	Wolfgang	Caltech	Max Williams
Kobayashi	Albert	IIT	Paul R. Trumpler
Krajcinovic	Dusan	Northwestern	George Herrmann
Kraynik	Andrew	Princeton	W. R. Schowalter
Ku	Yu	MIT	No advisor
Kyriakides	Stelios	Caltech	Charles D. Babcock
Lamb	J. Parker	Illinois	J. H. Bartlett
Landis	Chad	UC Santa Barbara	R. M. McMeeking
Leal	L. Gary	Stanford	Andreas Acrivos
Lee	Erastus	Stanford	Stephen P. Timoshenko
Leibovich	Sidney	Cornell	Geoffrey S. S. Ludford
Libby	Paul	Brooklyn Poly	Paul Lieber
Liepmann	Hans	ETH Zurich	Richard Bar

(continued)

(continued)

Last name	First name	Ph.D. University	Advisor
Liu	Wing Kam	Caltech	Thomas J. R. Hughes
Longmire	Ellen	Stanford	John K. Eaton
Lumley	John	Johns Hopkins	Stan Corrsin
Markovitz	Hershel	Columbia	G. B. Kimball
Marsden	Jerrold	Princeton	Arthur Strong Wrightman
Martin	Monroe	Johns Hopkins	Aurel Wintner
McDowell	David	Illinois	Darrell Socie
McKinley	Gareth	MIT	R. Brown & R. Armstrong
McMeeking	Robert	Brown	James R. Rice
Mear	Mark	Harvard	John Hutchinson
Mei	Chiang	Caltech	Theodore Y. Wu
Miklowitz	Julius	Michigan	F. L. Everett
Miles	John	Caltech	Homer J. Stewart
Mindlin	Raymond	Columbia	no advisor
Moin	Parviz	Stanford	W. C. Reynolds
Morkovin	Mark V.	Wisconsin	Ivan S. Sokolnikoff
Morris	Jeffrey	Caltech	John Brady
Murman	Earll	Princeton	Seymour Bogdonoff
Naghdi	Paul	Michigan	Paul F. Chenea
Newmark	Nathan	Illinois	Hardy Cross
Nordgren	Ronald	Cal Berkeley	Paul M. Naghdi
Norris	Andrew	Northwestern	Jan D. Achenbach
Noton	Bryan	no Ph.D.	no advisor
Ochoa	Ozden	Texas A & M	Thomas J. Kozik
Oden	J. Tinsley	Oklahoma State	Jan Tuma
Pao	Yih-Hsing	Columbia	Raymond D. Mindlin
Paris	Paul	Lehigh	Ferdinand P. Beer
Peercy	Paul	Wisconsin	Richard N. Dexter
Peskin	R.	Pittsburgh	Ludwig Prandtl
Pierce	Allan	MIT	Laszlo Tisza
Plunkett	Robert	MIT	J. P. den Hartog
Popov	Egor	Stanford	Stephen P. Timoshenko
Prager	William	Darmstadt	Wilhelm Schlink
Prosperetti	Andrea	Caltech	Milton Plesset
Ramberg	Walter	ETH Munich	Jonathan Zenneck
Ravi-Chandar	Krishnaswamy	Caltech	Wolfgang G. Knauss
Ravichandran	Guruswami	Brown	Rodney J. Clifton
Reissner	Eric	MIT	Dirk J. Struik
Rice	James	Lehigh	Ferdinand P. Beer
Rivlin	Ronald	Cambridge	Sc. D. no advisor
Roshko	Anatol	Caltech	Hans W. Liepmann
Rouse	Hunter	ETH Karlsruhe	Theodor Rehbock

(continued)

(continued)

Last name	First name	Ph.D. University	Advisor
Sackman	Jerome	Columbia	Hans Bleich
Saffman	Philip	Cambridge	George K. Batchelor
Salvadori	Mario	Univ. Rome	Unkown
Schowalter	William	Illinois	H. Fraser Johnstone
Schubauer	Galen	Johns Hopkins	Hugh L. Dryden
Seeger	Raymond	Yale	Leigh Page & R. Bruce Lindsay
Shaqfeh	Eric	Stanford	Andreas Acrivos
Shield	Richard	Durham	Albert E. Green
Shukla	Arun	Maryland	James W. Dally & George Irwin
Skalak	Richard	Columbia	H. H. Bleich
Smits	Alexander	Melbourne	A. E. Perry
Smoluchowski	R.	Groningen	Dirk Coster
Sottos	Nancy	Delaware	Roy McCullough
Spanos	Pol	Caltech	Wilfred D. Iwan
Stebe	Kathleen	City U. New York	Charles Maldarelli
Stone	Howard	Caltech	L. Gary Leal
Sture	Stein	Colorado	Hon-Yim Ko
Suo	Zhigang	Harvard	John Hutchinson
Tadmor	Eitan	Tel-Aviv Univ.	Saul S. Abarbanel
Taylor	Charles	Illinois	Henry Langhaar
Timoshenko	Stephen	Kiev Polytechnic	Ludwig Prandtl & Viktor Kirpichov
Todd	Judith	Cambridge	James Charles
Toupin	Richard	Syracuse	Melvin Lax
Troian	Sandra	Cornell	N. D. Mermin
Tulin	Marshall	MIT—(MS)	Holt Ashley
Van Dyke	Milton	Caltech	Paco Lagerstrom
von Kármán	Theodore	Göttingen	Ludwig Prandtl
von Mises	Richard	Vienna	Gerog Hamel
Vorus	William	Michigan	Horst Nowacki & T. Francis Oglivie
Wahl	Arthur	Pittsburgh	Arpad Nadai
Wallace	James	Oxford	Ray Franklin
Wan	Frederic	MIT	Eric Reissner
Wei	Robert	Princeton	Shao Lee Soo
Widnall	Sheila	MIT	Holt Ashley
Young	Y. L.	Texas	Spyros Kinnas

Appendix J: U.S. National Congress Proceedings

1951	1st U.S. Congress, Illinois Institute of Technology, Chicago, IL, June 11–16, 1951. General Chairman: L. H. Donnell. *Proceedings of the First U.S. National Congress of Applied Mechanics,* edited by E. Sternberg, ASME, 1952
1954	2nd U.S. Congress, University of Michigan, Ann Arbor, MI, June 14–18, 1954. General Chairman: E. L. Eriksen. *Proceedings of the Second U.S. National Congress of Applied Mechanics,* edited by P. M. Naghdi, ASME, 1955
1958	3rd U.S. Congress, Brown University, Providence, RI, June 11–14, 1958. General Chairman: W. Prager. *Proceedings of the Third U.S. National Congress of Applied Mechanics,* edited by R. M. Haythornthwaite and L. H. Donnell
1962	4th U.S. Congress, University of California, Berkeley, CA, June 18–21, 1962. General Chairman: W. W. Soroka. *Proceedings of the Fourth U.S. National Congress of Applied Mechanics* (two volumes), edited by R. M. Rosenberg, ASME, 1962
1966	5th U.S. Congress, University of Minnesota, Minneapolis, MN, June 14–17, 1966. General Chairman: B. J. Lazan. *Proceedings of the Fifth U.S. National Congress of Applied Mechanics,* edited by L. E. Goodman, ASME, 1966
1970	6th U.S. Congress, Harvard University, Cambridge, MA, June 15–19, 1970. General Chairman: H. W. Emmons. *Proceedings of the Sixth U.S. National Congress of Applied Mechanics,* edited by G. F. Carrier, ASME, 1970
1974	7th U.S. Congress, University of Colorado, Boulder, CO, June 3–7, 1974. General Chairman: F. Essenburg. *Proceedings of the Seventh U.S. National Congress of Applied Mechanics,* edited by S. K. Datta, ASME, 1974
1978	8th U.S. Congress, University of California Los Angeles (Los Angeles, CA), June 25–30, 1978. General Chairman: J. D. Cole. *Proceedings of the Eighth U.S. National Congress of Applied Mechanics,* edited by R. E. Kelly, Western. Periodicals Co., North Hollywood, CA, 1979
1982	9th U.S. Congress, Cornell University, Ithaca, NY, June 21–25, 1982, General Chairman: Y. H. Pao. *Proceedings of the Ninth U.S. National Congress of Applied Mechanics,* edited by Y. H. Pao. ASME, 1982
1986	10th U.S. Congress, University of Texas, Austin, TX, General Chairman: E. B. Becker. *Proceedings of the Tenth U.S. National Congress of Applied Mechanics* edited by L. J. Parker. ASME, 1987

(continued)

© Springer International Publishing Switzerland 2016
C.T. Herakovich, *Mechanics IUTAM USNC/TAM*,
DOI 10.1007/978-3-319-32312-1

(continued)

1990	11th U.S. Congress, University of Arizona, Tucson, AZ, May 21–25, 1990. General Chairman: C. F. Chen. Mechanics 1990: *Proceedings of the Eleventh U.S. National Congress of Applied Mechanics*, edited by C. F. Chen. ASME, 1990
1994	12th U.S. Congress, University of Washington, Seattle, WA, June 27–July 1, 1994. General Chair: A. S. Kobayashi. *Proceedings of the Twelfth U.S. National Congress of Applied Mechanics*, edited by A. S. Kobayashi
1998	13th U.S. Congress, University of Florida, Gainesville, FL, June 21–26, 1998. *Proceedings of the Thirteenth U.S. National Congress of Applied Mechanics*, Abstracts, edited by Martin A. Eisenberg
2002	14th U.S. Congress, Virginia Polytechnic Institute and State University, Blacksburg, VA, June 23–28, 2002. *Contemporary Research in Theoretical and Applied Mechanics, Proceedings of the Fourteenth U.S. National Congress of Applied Mechanics (ISBN 0-9721257-0-1)*, edited by R. C. Batra and E. G. Henneke. Virginia Tech
2006	15th U.S. Congress, University of Colorado, Boulder, CO, June 25–30, 2006. *Proceedings of the Fifteenth U.S. National Congress of Applied Mechanics (Proceedings CD)*, edited by Co-Chairs Stein Sture and Tom Geers. University of Colorado
2010	16th U.S. Congress, Penn State University, University Park, PA, June 27–July 2, 2010. *Proceedings of the Sixteenth U.S. National Congress of Theoretical and Applied Mechanics (CD-ROM)*, edited by Judy A. Todd and Charles E. Bakis, Co-Chairs
2014	17th U.S. Congress, Michigan State University, East Lansing, MI, June 15–20, 2014, Congress Co-Chaired by J. F. Foss and T. J. Pence, *Proceedings of the Seventeenth U.S. National Congress of Applied Mechanics,* edited by J. F. Foss and T. J. Pence (to be published)
2018	18th U.S. Congress, Northwestern University, Evanston, IL General Chairman: Jianmin Qu (Future congress)

Appendix K: U.S. Chaired IUTAM Symposia & Summer Schools

Year	Location, Title, Chairs, Dates, Publisher, ISBN/Ref., IUTAM Reference Number
1950	Hershey, PA, Plastic Flow & Deformation within the Earth, Sep. 12–14, T. Am. Geo. Union, v32, #4,497, 1951, 50-2
1957	Cambridge, MA, Cosmical Gas Dynamics, June 24–29, AIP, 57-1
1959	Ithaca, NY, Fluid Mechanics in the Ionosphere, R. Bolgiano, Jr., July 9–15, J. Geophy Res., V64 (dec. 1959) 2037, 59-1
1960	Washington, DC, Magnetohydrodynamics, Jan. 17–23, RMP, AIP, 60-1
1963	Providence, RI, Stress Waves in Anelastic Solids, H. Kolsky & W. Prager, Apr. 3–5, Springer-Verlag, 1964, 63-1
1964	Ann Arbor, MI, Concentrated Vortex Motions in Fluids, D. Kuchemann, July 6–11, J. Fluid Mechanics, 21:1:1–20, 1965, 64-2
1968	Monterey, CA, High-speed Computing in Fluid Dynamics, F. N. Frenkiel; K. Stewartson, Aug. 18–24, Phy Fluids, Supplement II, V129, #12, Part II, 1969, 68-2
1969	Cambridge, MA, Electrohydrodynamics, J. R. Melcher, Mar. 31–Apr. 2, J. Fluid Mechanics, Supp. II, V129, #12, Part II (1969), 69-2
1973	Charlottesville, VA, Turbulent Diffusion in Environmental Pollution, F. N. Frenkiel; R. E. Munn, Apr. 8–14, Ad Geophysics, V18A & 18B, Acad. Press (1974), 73-1
1974	Cambridge, MA, Buckling of Structures, B. Budiansky, June 17–21, Springer-Verlag, 1976, 74-1
1976	Washington, DC, Structure of Turbulence & Drag Reduction, F. N. Frenkiel, M. T. Landahl & J. L. Lumley, June 7–12, AIP, Phy. Fluids Part II, V20, #10, 1977, 76-1
1977	Evanston, IL, Modern Problems in Elastic Wave Propagation, J. Miklowitz & J. D. Achenbach, Sep. 11–15, Wiley, 1978, 77-3
1978	Evanston, IL, Variational Methods in the Mechanics of Solids, S. Nemat-Nasser, Sept. 11–13, Pergamon, 1980, 78-5
1980	Bethlehem, PA, Finite Elasticity, D. E. Carlson & R. T. Shield, Aug. 11–15, Martinus Nijhoff, 9024726298, 80-3
1981	Pasadena, CA, Mechanics & Physics of Bubbles in Liquids, L. van Wijngaarden, June 15–19, Martinus Nijhoff, 90-247-2625-5, 81-3
1982	Blacksburg, VA, Mechanics of Composite Materials, Z. Hashin & C. T. Herakovich, Aug. 16–19, Pergamon, 1983, 82-6

(continued)

© Springer International Publishing Switzerland 2016
C.T. Herakovich, *Mechanics IUTAM USNC/TAM*,
DOI 10.1007/978-3-319-32312-1

(continued)

1983	Evanston, IL, Mechanics of Geomaterial: Rocks, Concretes, Soils, Z. P. Bazant, Sep. 11–15, Wiley, 1985, 83-7
1987	San Antonio, TX, Advanced Boundary Element Methods: Solids & Fluids, T. A. Cruse, Apr. 13–16, Springer-V, 3-540-17454-0, 87-2
1988	Pasadena, CA, Recent Advances in Nonlinear Fracture Mechanics, W. G. Knauss & A. J. Rosakis, Mar. 14–16, IJF (V42,#1–4), 1990, 07923-0658-9, 88-2
1989	Boulder, CO, Elastic Wave Propagation & Ultrasonic Evaluation, S. K. Datta, J. D. Achenbach, Y. S. Rajapakse, Jul. 30–Aug. 3, North-Holland, 0-444-87485-2, 89-3
1990	Troy, NY, Inelastic Deformation of Composite Materials, G. J. Dvorak, May 29–June 1, Springer-Verlag, 0-387-97458-X, 90-3
1990	La Jolla, CA, Fluid Mechanics of Stirring & Mixing, H. Aref & J. M. Ottino, Aug. 20–24, Phy. Fluids A, V3 (1991), 1009–1469, ISSN 0899-8213, 90-6
1991	Stanford, CA, Mechanics of Fluidized Beds, G. M. Homsy, R. Jackson & J. R. Grace, July 1–4, J. Fluid Mechanics, V236, 477–495, 91-4
1991	Blacksburg, VA, Local Mechanics Concepts for Composite Material Systems, J. N. Reddy & K. L. Reifsnider, Oct. 27–31, Springer-Verlag, 3-540-55547-1, 91-10
1993	San Antonio, TX, Probabilistic Structural Mechanics: Adv. Structural Reliability Methods, P. D. Spanos & Y. -T. Wu, June 7–10, Springer-Verlag, 3-540-57709-2, 93-1
1993	Providence, RI, Computational Mechanics & Materials, M. Ortiz & C. F. Shih, June 15–18, J. Model Simulation in Matls Sci & Engr., V2, #3A, 421 (1994) ISSN 0965-0393, 93-2
1993	Potsdam, NY, Nonlinear Instability of Nonparallel Flows, S. P. Lin, W. R. C. Phillips, D. T. Valentine, Sep. 12–14, Springer-Verlag, 3-540-57679-7, 93-4
1997	Ithaca, NY, New Applications of Nonlinear & Chaotic Dynamics in Mechanics, F. C. Moon, July 27–Aug. 1, Kluwer, 0-7923-5276-9, 97-10
1997	Boulder, CO, Computational Methods for Unbounded Domains, T. L. Geers, July 27–31, Kluwer, 0-7923-5266-1, 97-11
1998	Boulder, CO, Developments in Geophysical Turbulence, R. M. Kerr & Y. Kimura, June 16–19, Kluwer, 0-7923-6673-5, 98-3
1998	Stanford, CA, Viscoelastic Fluid Mechanics, G. M. Homsy & E. S. G. Shaqfeh, June 21–25, J. Non-Newtonian Fluid Mechanics, V82 (1999), 127–457, 98-4
1999	Cape May, NJ, Segregation in Granular Flows, A. D. Rosata & D. L. Blackmore, June 5–10, Kluwer, 0-7923-6547-X, 99-5
1999	Notre Dame, IN, Nonlinear Wave Behavior in MultiPhase Flow, H. C. Chang, July 7–9, Kluwer, 0-7923-6454-6, 99-6
1999	Sedona, AZ, Laminar-Turbulent Transition, H. Fasel & W. S. Saric, Sep. 12–18, Springer-Verlag, 3-540-67947-2, 99-8
2000	Fairbanks, AK, Scaling Laws in Ice Mechanics & Ice Dynamics, J. P. Dempsey & H. H. Shen, June 13–16, Kluwer, 1-4020-0171-1, 00-2a
2001	Austin, TX, Material Instabilities & the Effect of Microstructure, S. Kyriakides & N. Triantafyllidis, May 7–10, Elsevier, IJSS #39, 2002, 01-2
2002	Austin, TX, Micromechanics of Fluid Suspensions and Composites, R. T. Bonnecaze, Z. Hashin, G. J. Rodin, April 3–5, Kluwer, PhilTrans: Math, Physical & Engr. Sci. May 2003, 02-1
2002	Urbana-Champaign, IL, Nonlinear Stochastic Dynamics, N. Sri Namachchivaya & Y. K. Lin, Aug. 25–31, Kluwer, 1-4020-1471-6, 02-6

(continued)

2002	Princeton, NJ, Reynolds Number Scaling in Turbulent Flow, A. J. Smits, Sep. 11–13, Kluwer, 1-4020-1775-8, 02-8
2003	Rutgers, NJ, Integrated Modeling of Fully Coupled Fluid-Structure Interactions, H. Benaroya, June 2–6, Kluwer, 1-4020-1806-1, 03-2
2004	Argonne, IL, Recent Advances in Disperse Multiphase Flow Simulation, S. Balachandar & A. Prosperetti, Oct. 4–7, Springer, 1-4020-4976-7, 04-7
2008	Cape Cod, MA, Cellular, Molecular & Tissue Mechanics, K. Garikipati & E. Arruda, June 18–21, Springer, 10-9048133475, 08-3
2009	Austin, TX, Dynamic Fracture & Fragmentation, K. Ravi-Chandar, March 8–13, 09-1
2011	Shalimar, FL, Linking Scales in Computations: From Microstructure to Macro-scale Properties, O. Cazacu, May 16–19, 08-09
2011	Stanford, CA, Computer Models In Biomechanics: From Nano to Macro, E. Kuhl, Aug. 29–Sep. 2, 08-06
2011	Austin, TX, Mechanics of Liquid and Solid Foams, S. Kyriakides & A. M. Kraynik, May 8–11, 08-05
2014	Evanston, IL, Connecting Multiscale Mechanics to Complex Material Design, W. K. Liu, May 14–16, 12-10

Summer schools hosted in the United States

2010	Evanston, IL, Mechanics of Soft Materials, Y. Huang & W. K. Liu, May 10–12.
2012	West Lafayette, IN, Biomechanics of Tissue and Tissue-Cell Interaction, T. Siegmund, June 5–8.

IUTAM symposia held outside the United States with a Co-Chair from the U.S.

1980	Dourdan, France, Three-Dimensional Constitutive Relations & Ductile Fracture, S. Nemat-Nasser, June 2–5, North-Holland, 10-0444861084, 80-2
1991	Beijing, China, Constitutive Relations for Finite Deformations of Polycrystalline Metals, R. Wang & D. C. Drucker, July 22–25, Springer-Verlag, 3-540-55128-X, 91-6
1996	Marseille, France, Variable Density Low Speed Turbulent Flows, L. Fulachier, J. L. Lumley & F. Anselmet, July 7–10, Kluwer, 0-7923-4602-5, 96-3
1997	Cairo, Egypt, Transformation Problems in Composite and Active Materials, Y. A. Bahei-El-Din & G. J. Dvorak, March 9–12, Kluwer, 0-7923-5122-3, 97-2
1999	Cracow, Poland, Advanced Mathematical & Computational Mechanics Aspects of the Boundary Element Method, T. Burczynski & T. A. Cruse, May 30–June 3, Kluwer, 0-7923-7081-3, 99-4
2006	Guanajuato, Mexico, Interactions for Dispersed Systems in Newtonian & Viscoelastic Fluids, G. M. Homsy & J. R. Zenit, March 26–31, J. Physics Fluids, V18, 121501-1, 2006, 06-2
2006	Lyon, France, Discretization Methods for Evolving Discontinuities, A. Combescure, T. Belytschko & R. de Borst, Sept. 4–7, Springer-Verlag, 978-1-4020-6529-3, 06-5
2007	Istanbul, Turkey, Recent Advances in Multiphase Flows: Numerical & Experimental, A. Acrivos & C. F. Delale, June 11–14, J. Physics Fluids, 07-2
2008	Lyngby, Denmark, 150 Years of Vortex Dynamics, H. Aref, Oct. 12–17, Springer, 978-90-481-8583-2, 08-5
2011	Paris, France, Full-field Measurements and Identification in Solid Mechanics, F. Hild & H. Espinosa

Appendix L: ICTAM Congresses

International Congresses of Applied Mechanics

International Congresses on mechanics have been held since 1924. Since 1926, they have been held at four-year intervals except for a period during the Second World War. Prior to 1964 the organization of the International Congress of Applied Mechanics was supervised by the *International Committee for the Congresses of Applied Mechanics*. The organization of each congress was under the overall supervision of a Scientific Committee appointed by the International Committee. The organizational details of each congress were entrusted to a local Organizing Committee that also undertook the publication of the proceedings. Unfortunately, there is no central office from which proceedings may be ordered; for each volume, a request must be made to the publishers of that particular congress. A complete listing of congresses, including information on the proceedings of some of the congresses, is provided in Appendix L.

1924	1st International Congress, Delft (Netherlands), 22–26 April 1924. Proceedings of the First International Congress for Applied Mechanics, Delft 1924, edited by C. B. Biezeno and J. M. Burgers (one vol.). Technische Boekhandel en Drukkerij J. Waltman Jr., Delft, 1925
1926	2nd International Congress, Zürich (Switzerland), 12–17 September 1926. Verhandlungen—Comptes rendus—*Proceedings of the 2nd International Congress for Applied Mechanics*, Zürich, 12–17 September 1926, Herausgegeben von E. Meissner (one vol.). Orell Füssli Verlag, Zürich und Leipzig, 1927
1930	3rd International Congress, Stockholm (Sweden), 24–29 August 1930. *Verhandlungen—Compte rendus—Proceedings of the 3rd International Congress for Applied Mechanics*, edited by A. C. W. Oseen und W. Weibull (3 vol.). AB. Sveriges Litografiska Tryckerier, Stockholm, 1931

(continued)

(continued)

1934	4th International Congress, Cambridge (UK), 3–9 July 1934. *Proceedings of the Fourth International Congress for Applied Mechanics*, Cambridge, UK, 3–9 July, 1934 (one vol.). University Press, Cambridge (UK), 1935
1938	5th International Congress, Cambridge (Massachusetts, USA), 12–16 September 1938. *Proceedings of the Fifth International Congress for Applied Mechanics*, held at Harvard University and the Massachusetts Institute of Technology, Cambridge, Massachusetts, September 12–16, 1938, edited by J. P. den Hartog and H. Peters (one vol.), John Wiley and Sons, Inc., New York (USA), and Chapman and Hall Ltd., London (UK), 1939
1946	6th International Congress, Paris (France), 22–29 September 1946
1948	7th International Congress, London (UK), 5–11 September 1948. *Proceedings of the Seventh International Congress for Applied Mechanics*, 1948, published by the Organizing Committee (Introduction, Vol. I, Vol. II—Parts 1 and 2, Vol. III, Vol. IV)
1952	8th International Congress, Istanbul (Turkey), 20–28 August 1952
1956	9th International Congress, Brussels (Belgium), 5–13 September 1956
1960	10th International Congress, Stresa (Italy), 31 August–7 September 1960. Proceedings published by the Consiglio Nazionale delle Ricerche, Piazelle delle Science 7, Roma (Italia), printed by Elsevier Publishing Company, Amsterdam-New York, 1962
1964	11th International Congress (ICTAM), Munich (Germany), 30 August–5 September 1964
1968	12th International Congress (ICTAM), Stanford, Cal. (USA), 26–31 August 1968. *The Proceedings*, edited by M. Hetényi and W. G. Vincenti, published by Springer-Verlag, Berlin (Germany), 1969
1972	13th International Congress (ICTAM), Moscow (USSR), 21–26 August 1972. *The Proceedings*, edited by E. Becker and G. K. Mikhailov, published by Springer-Verlag, Berlin (Germany), 1973
1976	14th International Congress (ICTAM), Delft (Netherlands), 30 August–4 September 1976. *The Proceedings*, edited by W. T. Koiter, published by North-Holland Publishing Company, Amsterdam-New York-Oxford, 1976, 1977
1980	15th International Congress (ICTAM), Toronto (Canada), 17–23 August 1980. *The Proceedings*, edited by F. P. J. Rimrott and B. Tabarrok, published by North-Holland Publishing Company, Amsterdam-New York-Oxford, 1980
1984	16th International Congress (ICTAM), Lyngby (Denmark), 19–25 August 1984. *The Proceedings*, edited by F. I. Niordson and N. Olhoff, published by Elsevier Science Publishers (North-Holland), Amsterdam, 1985
1988	17th International Congress (ICTAM), Grenoble (France), 21–27 August 1988. *The Proceedings*, edited by P. Germain, M. Piau and D. Caillerie, published by North-Holland, Elsevier Science Publishers, Amsterdam, 1989. ISBN 0-444-87302-3
1992	18th International Congress (ICTAM), Haifa (Israel), 22–28 August 1992. *The Proceedings*, edited by S. R. Bodner, J. Singer, A. Solan and Z. Hashin, published by Elsevier Science Publishers, Amsterdam, 1993. ISBN 0-444-88889-6
1996	19th International Congress (ICTAM), Kyoto (Japan), 25–31 August 1996. *The Proceedings*, edited by T. Tatsumi, E. Watanabe, T. Kambe, published by Elsevier Science Publishers, Amsterdam, 1997. ISBN 0-444 82446-4

(continued)

(continued)

2000	20th International Congress (ICTAM), Chicago (USA), 27 August–2 September 2000. *The Proceedings*, entitled *"Mechanics for a New Millenium"*, edited by H. Aref and J. W. Phillips, published by Kluwer Academic Publishers, Dordrecht, The Netherlands, 2001. ISBN 0-7923-7156-9
2004	21st International Congress (ICTAM), Warsaw (Poland), 15–21 August 2004. *The Proceedings*, entitled *"Mechanics of the twenty-first century"*, edited by W. Gutkowski and T. A. Kowaleski, published by Springer, Dordrecht, The Netherlands, 2005. ISBN 1-4020-3456-3
2008	22nd International Congress (ICTAM), Adelaide (Australia), 24–29 August 2008. *The Proceedings*, entitled *"Mechanics Down Under"* and edited by J. Denier and M. Finn, published by Springer, 2013. CXXII, 305 p. 139 illus
2012	23nd International Congress (ICTAM), Beijing (China), 19–24 August 2012. *The Proceedings* edited by Yilong Bai, Jianxian Wang and Daining Fang. Available on CD
2016	24nd International Congress (ICTAM), Montreal (Canada), (to be held on August 21–26, 2016)

References

Anderson, J. D. Jr.: Ludwig Prandtl's boundary layer. Phys. Today, 42–48 (2005, December)

Beer, F.P., Johnston Jr., E.R.: Vector Mechanics for Engineers: Statics and Dynamics. McGraw-Hill, New York (1962b)

Bernoulli, D.: Hydrodynamica, Argentorati (1738)

Besseling, J.F., van der Heijden, A.M.A.: Trends in Solid Mechanics 1979. In: Proceedings of the Symposium dedicated to the 65th Birthday of W. T. Koiter. Koiter, Forty Years in retrospect, the Bitter and the Sweet, p. 237. Delft University Press, (1979)

Boley, B.A., Weiner, J.H.: Theory of Thermal Stresses. Wiley, New York (1960)

Cauchy, A.-L.: De la pression ou tension dans un corps solide, [On the pressure or tension in a solid body]. Exercices de Mathématiques. 2, 42 (1827)

Christensen, R.M.: Mechanics of Composite Materials. Wiley, New York (1979)

Cottrell, J.A., Hughes, T.J.R., Bazilevs, Y.: Isogeometric Analysis. Wiley, Chichester (2009)

D'Alembert, J.-B.R.: Traité de Dynamique. Paris, (1743) (reprint Johnson Reprint Corp.,1968)

Den Hartog, J.P.: Mechanical Vibrations, 4th edn. McGraw-Hill, New York (1956)

Dugas, R.: A History of Mechanics 1988, translation by J. H. Miller. Dover, New York (1955)

Einstein, A.: Does the inertia of a body depend upon its energy content? Ann. Phys. 18, 639 (1905a)

Einstein, A.: On a heuristic viewpoint concerning the production and transformation of light. Ann. Phys. 17, 132 (1905b)

Einstein, A.: On the electrodynamics of moving bodies. Ann. Phys. 17, 891–921 (1905c)

Einstein, A.: On the motion of small particles suspended in a stationary liquid, as required by the molecular kinetic theory of heat. Ann. Phys. 17, 549–560 (1905d)

Euler, L.: Mechanica (1736), in Opera Omnia, Birkhauser, 1911

Freund, L.B., Suresh, S.: Thin Film Materials. Cambridge University Press, Freund publications (2003)

Freund, L.B.: Dynamic Fracture Mechanics. Cambridge University Press, Freund publications (1990)

Galileo, G.: Discourses and Mathematical Demonstrations Relating to Two New Sciences, (Discorsi e dimostrazioni matematiche, intorno à due nuove scienze). (1638) House of Elzevir

Gorn, M.H.: The Universal Man, Theodore von Kármán's Life in Aeronautics. Smithsonian Institution Press, Washington, DC (1992)

Green, G.: An Essay on the Application of Mathematical Analysis to the Theories of Electricity and Magnetism, Nottingham (1828) Crelle's Journal

Green, G.: On the laws of reflexion and refraction of light at the common surface of two non-crystallized media. Camb. Phil. Soc. Trans. 7, 1–24 (1839)

© Springer International Publishing Switzerland 2016
C.T. Herakovich, *Mechanics IUTAM USNC/TAM*,
DOI 10.1007/978-3-319-32312-1

Hardie, R.P., Gaye, R.K.: Physica. In: Ross, W.D. (ed.) The Works of Aristotle v. 2. Clarendon, Oxford (1930)

Herakovich, C.T.: Mechanics of Fibrous Composites. Wiley, New York (1998)

Kármán, T.V., Edson, L.: The Wind and Beyond—Theodore von Kármán Pioneer in Aviation and Pathfinder in Space (Autobiography). Little Brown, Boston (1967)

Lagrange, J.L.: Mécanique Analytique, (1788), and later as 4. ed., vol. 2. Gauthier-Villars et fils, Paris (1888–1889)

Lamb, H.: Hydrodynamics. Cambridge University Press, Cambridge (1932)

Laplace, P.S.: Traité de méchanique céleste, vol. 5. Duprat and Bachelier, Paris (1799–1825)

Lekhnitskii, S.G.: Theory of Elasticity of an Anisotropic Body (English Translation by P. Fern, Holden-Day). Mir, Moscow (1950)

Love, A.E.H.: A Treatise on the Mathematical Theory of Elasticity. Cambridge University Press, Cambridge (1927)

Love, A.E.H.: Some Problems of Geodynamics. Cambridge University Press, Cambridge (1911)

Millikan, R.A.: The Autobiography of Robert A. Millikan. Prentice-Hall, New York (1950)

Mises, R.V.: Mathematical Theory of Compressible Fluid Flow. Academic, New York (1958b)

Muskhelishvili, N.I.: Some Basic Problems of the Mathematical Theory of Elasticity, Moscow, (1933) English translation, Noordhoff, 1977

Nadai, A.: Plasticity, a Mechanics of the Plastic State of Matter. McGraw-Hill, New York (1931)

Naghdi, P.: A brief history of the applied mechanics division of ASME. J. Appl. Mech. 46(4), 723–749 (1979b)

Navier, C.-L.: Lois de l'équilibre et du mouvement des corps solides élastiques, paper read to the Académie des Sciences, Paris (1821)

Newton, I.: Philosophiae Naturalis Principia Mathematica ("Mathematical Principles of Natural Philosophy") (1687) English translation by Andrew Motte, publisher: Benjamin Motte, 1729

Oden, J.T.: An Introduction to Mathematical Modelling: A Course in Mechanics. Wiley, New York (2011)

Poisson, S.D.: Treatise de mecanique. 2 vols. (1811) and (1833), Paris, Bachelier

Rayleigh, B., Strutt, J.W.: Theory of Sound, vol. 2. Macmillan, London (1878)

Sokolnikoff, I.S., Sokolnikoff, E.: Higher Mathematics for Engineers and Physicists. McGraw-Hill, New York (1941)

Sokolnikoff, I.S.: Mathematical Theory of Elasticity. McGraw-Hill, New York (1956b)

Stocks, J.L.: On the Heavens. Clarendon, Oxford (1922)

Timoshenko, S.P.: As I Remember. D. Van Nostrand, Princeton (1968)

Truesdell III, C.A.: A First Course in Rational Continuum Mechanics. General Concepts, vol. 1. Academic, New York (1977)

Winter, T.N.: The Mechanical Problems in the Corpus of Aristotle. University of Nebraska—Lincoln, Lincoln (2007). Natural Philosophy, 1687

Index

© Springer International Publishing Switzerland 2016
C.T. Herakovich, *Mechanics IUTAM USNC/TAM*,
DOI 10.1007/978-3-319-32312-1

Printed in the United States
By Bookmasters